Encyclopedia of Alternative and Renewable Energy: Conversion Techniques for Biofuels

Volume 15

Encyclopedia of Alternative and Renewable Energy: Conversion Techniques for Biofuels

Volume 15

Edited by **Robbie Larkin and David McCartney**

New York

Published by Callisto Reference,
106 Park Avenue, Suite 200,
New York, NY 10016, USA
www.callistoreference.com

**Encyclopedia of Alternative and Renewable Energy:
Conversion Techniques for Biofuels
Volume 15**
Edited by Robbie Larkin and David McCartney

International Standard Book Number: 978-1-63239-189-6 (Hardback)

Contents

Preface

It is often said that books are a boon to mankind. They document every progress and pass on the knowledge from one generation to the other. They play a crucial role in our lives. Thus I was both excited and nervous while editing this book. I was pleased by the thought of being able to make a mark but I was also nervous to do it right because the future of students depends upon it. Hence, I took a few months to research further into the discipline, revise my knowledge and also explore some more aspects. Post this process, I begun with the editing of this book.

This book provides analysis of latest conversion techniques for biofuels and describes production methods for gases and other products. It emphasizes on the recent advances in the production of liquid and gaseous biofuels that will be of interest to the chemical scientists and technologists.

I thank my publisher with all my heart for considering me worthy of this unparalleled opportunity and for showing unwavering faith in my skills. I would also like to thank the editorial team who worked closely with me at every step and contributed immensely towards the successful completion of this book. Last but not the least, I wish to thank my friends and colleagues for their support.

Editor

Gases and Other Products

Biofuels from Algae

Robert Diltz and Pratap Pullammanappallil

Additional information is available at the end of the chapter

1. Introduction

1.1. Current sources of biofuels

The United States, as well as numerous other countries throughout the world, is seeing a rapid rise in the amount of power and fuel required to maintain the current and future life-styles of its citizens. With the rapid increase in global consumerism and travel seen over the recent decades due to improvements in technology and the increase in international interactions, the demand for fuel is rapidly growing, as can be seen in Figure 1. Due to the world-wide demand for fuel, which currently is primarily fossil-derived, supplies are being strained and costs are rapidly rising. In order to satiate this rapid increase in demand and stem the shrinking supply, new alternative sources of fuel must be brought to the market that can be used to replace standard petroleum based fuels.

Currently, there are several sources of alternative fuels that can be used to replace or supplement traditional petroleum based fuels. Some of these sources include alternative fossil-derived sources such as coal, natural gas, and hydrogen derived from hydrocracking, while other sources come from more renewable sources such as biomass. Biomass has several advantages when it comes to fuels in that there are numerous sources such as terrestrially grown starch based or cellulosic material, waste derived material, or aquatic and marine based organisms, each of which has unique components and characteristics useful for fuel production.

Due to the structural variability of the various types of biomass available, a wide range of technologies can be used to convert the organic molecules into a useable form of fuel. As food substrates (such as carbon dioxide in autotrophic organisms or sugars in heterotrophic organisms) are metabolized, a range of cellular components are assembled to perform numerous duties to keep organisms alive and reproducing. Starches and celluloses are assembled from carbohydrates to provide rigid structural support in many woody biomasses as well as acting as a sugar storage method for quick conversion to a food source in times of famine. Proteins and amino acids are the building blocks of DNA structures and additional biomass. Lipids provide a highly energy dense storage system while also serving as a transport mechanism for several nutrients vital to metabolic activity. However, when broken down to the most basic levels, these organic compounds all contain energy which can be extracted through several methods. Table 1 shows a breakdown of some common algal biomass cellular components.

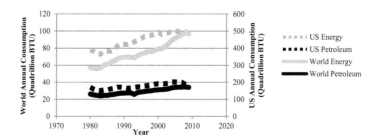

Figure 1. Annual Consumption of Total Energy and Petroleum in the United States and the World [1]

Species	Protein (%)	Carbohydrate (%)	Lipid (%)
Ankistrodesmus	36	24	31
Nitzschia	36	14	22
Chlorella	55	24	21
C. protothecoides	38	52	11
C. emersonii	32	41	29
C. vulgaris	29	51	18

Table 1. Variations in the chemical composition of selected algal species [2, 3]

Sources of biofuel currently being produced range in production rate from the laboratory scale through full scale implementation. Technologies to break down starches and cellulosic materials into sugars for subsequent conversion to bioalcohols has been extensively developed and scaled to produce billions of gallons per year to add into petroleum derived gasoline. Other structural components such as lipids have a high energy content to them and have characteristics that closely mimic petroleum diesel and kerosene, and thus, only require simple chemical reaction (i.e., transesterification) for use as a biofuel, and have been developed up to a quasi-large scale of volumetric output that can be seen in some regional market places, as well as in home production for personal use.

1.2. Aquatic biomass

In order to produce the vast amounts of fuel needed by the United States, and the rest of the world, there will be a demand for massive quantities of biomass to be grown. This could be problematic when using terrestrial biomass, since in most cases, growing plants would re-quire a switch from using land for food sources to energy sources. An alternative source of biomass, however, is available in the form of aquatic and marine species of biomass such as kelps, algae, and other types of water borne plants or bacteria. Aquatic and marine biomass (excluding bacteria) are typically plant-like in that they are autotrophic organisms that con-tain photosynthetic pigmentation, can utilize inorganic carbon for biomass development, and express molecular oxygen as a byproduct. However, these organisms do not suffer from the inherent liability of requiring fertile soil to grow, minimizing competition with the food supply chain. Also, as microscopic organisms, they do not require abundance of land to de-velop root systems and large floral brush in order to absorb sunlight and nutrients, and therefore, a much more effective utilization of space. With rapid growth rates that can typi-cally double in concentration in less than a day, it is possible to have daily harvests, creating a steady and abundant supply of biomass for harvesting. As such, marine and aquatic bio-mass can be a useful alternative source of biomass that can be used to produce a wide range of biofuels for commercial use, while avoiding several of the more common pitfalls associat-ed with more traditional sources of terrestrial biomass, and thus, will be the biomass focus of discussion for the remainder of this article.

Primarily, growth of algae for the production of oils and energy conversion has focused on microalgae, including species of diatoms and cyanobacteria (as opposed to macroalgae, such as seaweed), although some bacterial species (such as *Clostridium sp.)* have been demonstrat-ed for production of biologically derived hydrogen and methane [4]. To date, there have been numerous studies of algae and other water based biomass in order to identify strong candidates for biomass accumulation rates as well as lipid content for production of biodie-sel. Some strains are summarized for these characteristics in Table 2. There is also a wealth of microbial biomass resources available as a by-product of industrial activities such as sew-age treatment, brewing industries and food processing that could provide biomass or nu-trients for further microbial biomass growth [5, 6]. With this concept, it is feasible to use algae as a means for tertiary wastewater treatment in order to utilize trace nutrients such as phosphorous- and nitrogen-containing compounds, or can be used at industrial processes as a way to absorb carbon dioxide by entraining algal cultures to gaseous exhaust streams.

Growth of aquatic and marine biomass is not without challenges though. Maximum growth rates of the microorganisms typically occur under very specific conditions, and any variance on these conditions can cause substantial delays in biomass development. Also, open pond algal systems (which are common for algae production due to their ease of construction and inexpense) are susceptible to contamination from various airborne microorganisms that can decrease overall productivity. And of prime concern, is the ability to separate algae from water, which due to their very dilute nature, can be expensive and inefficient. Several meth-ods are used to do this, such as flocculation with chemicals (such as hydroxides or alum) or electric fields, filtration, centrifugation, or thermal drying, but each of these methods is not without bulky equipment, expensive materials, or long processing times.

2. Lipids and biodiesel

The diesel engine, created by Rudolph Diesel in 1893 as an alternative to steam engines, has seen a marked rise in use over the past decades as newer engines coming to market have become such cleaner combustors. Since the engines are so efficient, they are ideal for use in heavy transport such as rail and ship, but as technology and advances in fuel make the engine emissions cleaner, more and more small engine vehicles are coming to market in light trucks and passenger cars in the US and Europe as well as the rest of the world.

Species	Biomass Productivity (g/L/D)	Growth Rate (d⁻¹)	Biomass Conc. (g/L)	Lipid Content (% by dry weight)	Reference
Chlorella lutereorividis		0.55		28.5	[7]
Chlorella protothecides		1.32		31.2	[7]
Chlorella regularis		3		44.4	[7]
Chlorella vulgaris	1.9		1.9	53	[8]
Scenedesmus bijuga		6.1		35.2	[7]
Scenedesmus dimorphus		5.9		43.1	[7]
Scenedesmus obliquus		5.4		42.6	[7]
Dunaliella salina	0.3			35	[9]
Spirulina platensis	0.1				[9]
Tetraselmus chui	1			23.5	[10]
Botryoccocus braunii	10.8			25-75	[11]
Nannochloropsis sp	72			31-68	[11]
Nannochloropsis oculata	2.4			22.8	[10]
Phaeodactylum tricornutum		0.003		2.5	[12]

Table 2. Productivity of Selected Algal Species

Diesel engines have the ability to run on various sources of fuel. Originally the engine was tested using pure peanut oil and vegetable oil, though today, the engine is commonly run on fossil fuel based diesel fuel, a type of kerosene. To reduce the amount of petroleum based diesel being used in today's market several alternative types of fuel have been introduced that are compatible with these engines. Among the alternatives, generally seen are the lipid based straight vegetable oils and the modified biodiesels. Straight vegetable oil will burn without problem in diesel engines; however, preheating of the fuel is required in order to reduce viscosity to pumpable levels. Biodiesel fuels, which are generally from the same source of lipids as straight vegetable oils or algal oils, are a much better suited fuel because they match several of the same characteristics as modern diesel fuel, and thus, require little to no engine modifications or fuel pretreatment modifications.

2.1. Sources of lipids

Lipids are a general set of cellular components that are grouped together by the common trait that they are soluble in non-polar solvents. Throughout living organisms, there are several sources of lipids that play various roles in biochemical processes including energy storage and water insoluble nutrient transport across cell membranes that include neutral lipids, phospholipids, steroids, waxes, and carotenoids. Since lipids have a generally low oxygen and high carbon and hydrogen content, they are very energy dense molecules. This characteristic, along with their natural abundance and similarities with petroleum based fuels, make them ready targets for processing and use as a blend or replacement to traditional fuels.

Neutral lipids (commonly referred to as "fats"), which are widely regarded as one of the most common sources of lipids, and which has the highest potential for use as an alternative fuel, can be found in various forms throughout different organisms, and will be the primary topic of focus for this discussion. Most marine and aquatic biomass can store lipids within the cell that can range from a small fraction to upwards of 80% of the cellular weight. Due to this trait, research and production scale operations have been centered on utilizing aquatic biomass for lipid production and conversion to fuel with the remaining cellular components being recycled for mineral content or discarded.

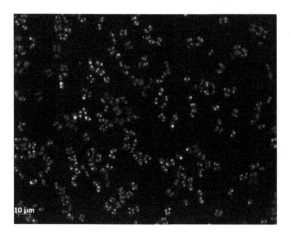

Figure 2. Nile Red Fluorescence Image of *Nitzchia sp.*

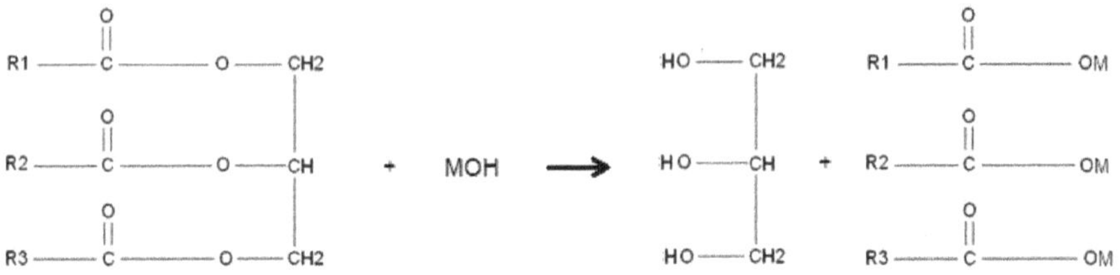

Scheme 1. Transesterification reaction schematic

Figure 2 shows an example of a marine diatom *Niztchia sp.* stained with Nile Red fluorescence stain (red color shows chlorophyll and yellow shows lipid fluorescence).

Neutral lipids consist of a glycerol molecule (a three carbon alcohol) and one to three fatty acids (referred to as mono-, di-, or tri- acylglycerols depending on number of fatty acids present) with the fatty acids being various carbon chain lengths and having various levels of unsaturation (unsaturated, mono-unsaturated, poly-unsaturated, etc.). Fatty tissues in animals serve as both an energy storage mechanism as well as a means of insulation against temperature extremes. Algae primarily store fats in the cell membrane to serve as an energy storage medium as well as a nutrient transport system to shuttle metabolites into and out of the cell. Several studies have been conducted to attempt to identify the distribution of fatty acids in algae and other aquatic biomass [13-15].

2.2. Ideal lipid characteristics for biodiesel

Biodiesel is produced primarily through the transesterification reaction of triglycerides and alcohol usually in the presence of a metal catalyst and can be visualized by the chemical reaction equation found in Scheme 1. where "R" groups are functional carbon chains varying in length and level of saturation and "M" is a metal, usually referring to sodium or potassium. The resultant glycerol that is produced is generally treated as a by-product and either sold for commodities use or burned to provide heating if necessary. This process is dependent on water content and pH, which dictates pre-processing demands in order to minimize the formation of soaps and maximize the production of wanted fatty acid ester compounds.

During this reaction the fatty acids tails are removed from the glycerol backbone leaving a glycerol molecule and one to three fatty acid esters (almost always either ethyl or methyl alcohol yielding a methyl or ethyl ester). These fatty acid methyl esters (FAME) or ethyl esters (FAEE) will vary in characteristics as a fuel based on carbon chain length as well as degree of unsaturation and location of unsaturated bonds. Some of the characteristics of biodiesel that are affected by fatty acid chemistry are viscosity, cloud point, and freezing point, among other factors important to engine performance. In general, there are several trade-offs that must be made with regards to saturation of fatty acids, branching of the fatty acid chain, and the carbon chain length, as each will have positive and negative attributes affecting fuel performance.

As the length of the molecule increases, the cetane number, and thus the heat of combustion, increases, this in turn decreases NO_x emissions. However, as the length of the fatty acid chain increases, the resultant biodiesel has increased viscosity leading to a pre-heating requirement. Also, as fatty acids become more branched there is a benefit of the gel point (the temperature at which the fuel becomes gel-like and has complications flowing through fuel lines) decreasing. The negative to higher branching is that the cetane number will decrease due to a more difficult combustion. As saturation of the fatty acid chain increases, there is a decrease in NO_x emissions and an improvement in fuel stability. As saturation increases, there is an increase in melting point and viscosity, both undesirable traits in a fuel.

Since there are so many trade-offs in the production of biodiesel, it is very difficult, if not impossible, to pick one ideal source of fatty acid for conversion to fuel. The multitude of climates across the globe will necessitate various traits in fuel such as the gel point, melting/freezing point, and oxidative stability. This leads to the argument of localized production of specific biomass sources that can be tailored to produce the types of lipids most suited to fuel that specific region, which will keep transportation costs down, as well as provide for the local economy. In following this method, there will be ample biomass produced to meet the specific needs of each climate, reducing environmental stresses that can occur due to overproduction for large scale purposes.

2.3. Enhancement of lipid production

Due to the various conditions that microorganisms grow and the constant flux of nutrients that can persist in nature, there are numerous types of lipids found that can change in concentration as the local environments evolve through typical ebbs and flows of materials. In response to these changes, microorganisms will change their cellular structures (i.e., lipid accumulation) by storing energy in various forms in order to utilize existing nutrients and energy to prepare for leaner conditions that may occur. In practical terms, this concept can be leveraged in order to produce high concentrations of intracellular lipids in marine and aquatic biomass in order to maximize the amount of lipids that can be harvested. Several studies have been conducted to determine what conditions affect the lipid composition and concentration of microorganisms. The more common techniques applied to increase the production of lipids from algae have through genetic manipulation [16], where genetic markers are manipulated that allow for increased lipid production to occur in the cell under normal conditions, by alteration of the cultivation conditions[17, 18], or by addition and manipulation of nutrients and chemicals added to the media [19]. By utilizing methods such as these, algal lipids can be increased by a substantial amount without increasing the footprint of required reactor space, nor greatly increasing the amount of time between harvests.

3. Synthetic fuels from biomass

3.1. Synthetic fuels

Unlike biofuels, which transform biological molecules into petroleum substitutes, synthetic fuels take a raw biological material, and through chemical processing, create compounds identical to petroleum fuel. This has a very distinct advantage over common biofuels in that there are no compatibility issues between the traditional fuels nor is there a need for any engine or fuel line modifications required. Synthetic fuels are usually made by utilizing a complex biological molecule and through thermal processing, break down the material into simple chemical building blocks (i.e., methane, carbon monoxide, hydrogen, etc.) and reform them into target chemicals. There are limitations with synthetic fuels production, especially when pertaining to production from aquatic and marine biomass where the water content is naturally higher than 99% by weight in its natural state, since initial breaking down of the products is usually through thermal processing that require dry or near dry conditions. However, since algae and aquatic biomass has such diverse characteristics and high cellular energy density, there is benefit for using either algae where the lipids have been extracted or whole algal cells as feedstock for these thermal synthetic fuel processes and thus can be considered as an option for production of synthetic fuels.

3.2. Methods for synthetic fuel production

There are three common methods for producing molecular precursors for synthetic fuels from biomass, and several variants of each method, dependent on the specific feedstock characteristic. These three methods are gasification, pyrolysis, and liquefaction. Pyrolysis and liquefaction will both produce of form of bio-oil that can be processed along with petroleum oil stocks and made into useful fuel products, while gasification will produce gaseous products such as carbon monoxide, methane, and hydrogen (commonly called syngas or synthesis gas in this process), and can be further refined directly to produce specific fuel molecules.

3.3. Gasification

Gasification is a process in which carbonaceous materials are exposed to heat and a sub-stoichiometric concentration of air to produce partially oxidized gaseous products that still have a high heating value with relatively lower concentrations of carbon dioxide due to limited oxygen [20]. Syngas can be catalytically reformed into a liquid fuel through the Fischer–Tropsch process, which converts carbon monoxide and hydrogen into long-chain hydrocarbons. By-products of the process include ash (formed from alkali-metal promoters present in the original reaction), char and tars that are created due to inefficiencies in mixing and heat distribution. This can be problematic when using water based biomass as the feedstock, since there will either be very high costs (in both energy and cost) to dry, or numerous unwanted products formed through side reactions. Three main types of gasification reactor are commonly used in industry: fixed bed, fluidized bed and moving bed. Each process has inherent advantages and drawbacks based on the complexity of the reactors, operating costs and product quality for use in the combustion of biomass. A more in-depth discussion of the design criteria and problems associated with using biomass as a fuel source for gasification reactors can be found in recent review articles [20, 21].

3.4. Liquefaction

Liquefaction is a process of converting biomass into a bio-oil in the presence of a solvent— usually water, an alcohol, or acetone—and a catalyst [22]. Liquefaction operates at milder temperatures than gasification, but requires higher pressures. Liquefaction can be indirect, wherein biomass is converted into gas and thence into liquid, or direct, in which biomass is converted directly into liquid fuel [23]. Bio-oils produced in direct liquefaction processes usually produce heavy oils with high heating values and value-added chemicals as by-products. Direct liquefaction also produces relatively little char compared to other thermochemical processes that do not utilize solvents. In addition, liquefaction has the advantage that the method is not hindered by the water content of the biomass, giving credence to utilizing this method for water based biomass. The use of water as a solvent can significantly reduce operating costs, and recent studies with sub-and super-critical water have demonstrated increased process productivity by overcoming heat-transfer limitations [24, 25]. Operating parameters and feed quality significantly influence the overall quality of the oil produced by these processes. A recent review presented an exhaustive comparison of the operational variables that affect the liquefaction of biomass and concluded that a well-defined temperature range is the most influential parameter for optimizing bio-oil yield and biomass conversion [22]. Similarly, catalyst choice can alter the heating value of the final liquefaction product and reduce the quantity of solid residue [25].

3.5. Pyrolysis

Pyrolysis is a process in which organic matter is exposed to heat and pressure in the absence of oxygen. The primary components of this process are syngas molecules like those found in gasification, as well as bio-oils and charred solid residues [26]. Pyrolysis methods are defined by the rate of heating, which directly affects the residence time of the reaction [27]. In slow pyrolysis, for example, the material is exposed to reactor conditions for five minutes; in fast pyrolysis, residence time is reduced to one to two minutes and in flash pyrolysis to less than five seconds. The residence time of the pyrolysis reaction greatly influences the composition of oils, gases and chars that are formed [28-30]. Several studies have been performed to identify the effect of operational variables— reactor conditions and variations in feedstock material —on the quality of the pyrolysis oils, gases, and chars [27, 30]. The oils typically produced during pyrolysis reactions are high in moisture content, and corrosive due to low pH. Pyrolysis of biomass is typically constrained by the high water content of the raw material, and current pyrolysis methods for biomass conversion have not reached the stage of commercial development. Ongoing research, however, aims at maximizing energy potential from biomass and optimizing conversion methods to achieve commercialization at marketable levels [31, 32].

4. Ethanol

Several species of cyanobacteria, including *Chlamydomonnas reinhardtii, Oscillatoria limosa, Microcystis PCC7806, Cyanothece PCC7822, Microcystis aeruginosa PCC7806 and Spirulina platensis* produce ethanol via an intracellular photosynthetic process. After selecting strains for ethanol, salt and pH tolerance, ethanol production can be enhanced through genetic modification [33]. These strains are long-lived and can be grown in closed photobioreactors to produce an ethanol containing algae slurry. This process for ethanol production from algae is currently being demonstrated by Algenol Biofuels [34-36]. The cyanobacteria are grown in flexible-film, polyethylene-based closed photobioreactors containing seawater or brackish water as medium. Industrial (or other waste) CO_2 is sparged into the bags to enhance growth of the microorganisms. Nutrients (primarily nitrogen and phosphorus) are supplied to sustain growth. At maturity, the microorganisms produce ethanol. The ethanol in the liquid phase will maintain an equilibrium with the ethanol-water in the vapor phase. The ethanol-water in vapor phase condenses along the walls of headspace which is collected by gravity for ethanol recovery. Algenol aims to produce 56,000 L of ethanol per hectare per year using 430 polyethylene bags established over a one hectare footprint each containing 4500 L of culture medium with a cyanobacteria concentration of 0.5 g/L. Unlike other algae derived biofuel processes, the algae are retained in the bags while the ethanol water condensate is removed for ethanol recovery. It is expected that the photobioreactors will be emptied once a year to replace the seawater, growth media and cyanobacteria.

The ethanol concentration in the algal cultures is expected to range between 0.5 and 5 % (w/w) depending on the ethanol tolerance levels of the strain and that of the condensate between 0.5 and 2% [36]. Since the maximum ethanol concentration is expected to be only 2 %, conventional distillation for ethanol recovery will not be energy efficient. A vapor compression steam stripping (VCSS) process is being developed to concentrate the ethanol to 5-30 % (w/w) range. VCSS is a highly heat integrated process that offers the potential for energy efficient separation even at low ethanol concentrations. This is then followed by a vapor compression distillation process to concentrate ethanol to an azeotropic 94% concentration. Life cycle energy requirements and greenhouse gas emissions for the process are dependent on the ethanol content of the condensate from the photobioreactors. Detailed analysis using process simulation software have shown that net life cycle energy consumption (excluding photosynthesis) is 0.55 down to 0.2 $MJ/MJ_{ethanol}$ and net life cycle greenhouse gas emissions is 29.8 down to 12.3 g $CO_2e/MJ_{ethanol}$ for ethanol concentrations ranging from 0.5 to 5% by weight [36]. Compared to gasoline these values represent a 67% and 87% reduction in the carbon footprint on an energy equivalent basis [36].

One of the technological challenges for this approach appears to be developing genetically engineered cyanobacterial strains that can tolerate high concentrations of ethanol. The ethanol concentration in the growth medium will affect the vapor phase ethanol content which in turn will affect the content of the condensate recovered from the photobioreactor. There is a dramatic increase in energy consumption in a conventional distillation process as ethanol content decreases below 7.5% (by volume). Energy required almost doubles when ethanol content decreases from 12% down to 5% (by volume).

Another challenge would be the economical disposal of spent algal cultures. Sterilization and inactivation of large volumes of biomass can involve extremely energy intensive unit operations like heating, or expensive processes like ultra violet treatment or chlorination.

5. Anaerobic digestion

Biogasification (or anaerobic digestion) is a biochemical process that converts organic matter to biogas (a mixture of methane, 50-70%, and balance carbon dioxide) under anaerobic conditions. Biogas can be used as a replacement for natural gas or it can be converted to electricity. The process is mediated by a mixed, undefined culture of microorganisms at near ambient conditions. Several terrestrial biomass feedstocks (agricultural residues, urban organic wastes, animal wastes and biofuel crops) have been anaerobically digested and commercial scale digesters exist for the biogasification of such feedstocks.

Anaerobic digestion offers several advantages over other biofuel production processes like ethanol fermentation or thermochemical conversion. The microbial consortia in an anaerobic digester are able to naturally secrete hydrolytic enzymes for the solubilization of macromolecules like carbohydrates, proteins and fats. Therefore, unlike in ethanol fermentation process there is no need to incorporate a pretreatment step to solubilize the macromolecules prior to fermentation. In addition, since the process is mediated by a mixed undefined culture, issues of maintaining inoculum (or culture) purity does not arise. Being a microbial process, there is no need to dewater the feedstock prior to processing unlike in thermochemical conversion where the feedstock is dried, to improve net energy yield. This is advantageous when it comes to processing aquatic biomass as these can be processed without dewatering. The anaerobic digestion process will also mineralize organic nitrogen and phosphorous, and these nutrients can be recycled for algae growth [37].

The process primarily takes place in four steps. A mixed undefined culture of mciroorganisms mediates hydrolysis, fermentation, acetogenensis and methanogenesis of the organic substrates as shown in Figure 3. During hydrolysis, the complex organic compounds are broken down into simpler, soluble compounds like sugars, amino acids and fatty acids. These soluble compounds are fermented to a mixture of volatile organic acids (VOA). The higher chain VOAs like propionic, butyric, and valeric acids are then converted to acetic acid in the acetogenesis step. Acetic acid is converted to methane during methanogenesis. Hydrogen and carbon dioxide are also liberated during fermentation and acetogenesis. A different group of methanogens converts hydrogen and carbon dioxide to methane. This mixed microbial culture thrives in the pH range of 6-8. Digestion can be performed either at mesophilic conditions (30 - 38ºC) or thermophilic conditions (49 - 57ºC).

Aquatic biomass – macrophytes [38], micro and macro algae, have all been tested as feedstock for biogasification. Microalgae have proportions of proteins (6–52%), lipids (7–23%)

and carbohydrates (5–23%) that are strongly dependent on the species and environmental conditions [39-41]. Compared with terrestrial plants microalgae have a higher proportion of proteins, which is characterized by a low carbon to nitrogen (C/N) ratio. The average C/N for freshwater microalgae is around 10.2 while it is 36 for terrestrial plants [40]. Usually the digestion of terrestrial plants is limited by nitrogen availability; however for microalgae this situation does not arise. Besides carbon, nitrogen and phosphorus, which are major components in microalgae composition, oligo nutrients such as iron, cobalt, zinc are also found [42]. These characteristics of microalgae make it a good feedstock for anaerobic digestion.

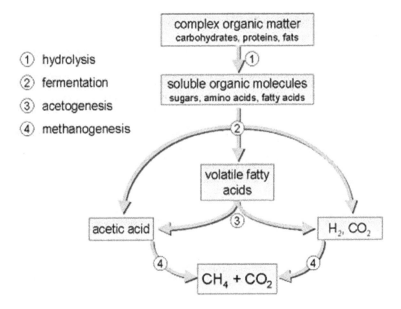

Figure 3. Pathways for mineralization of organic matter to biogas in an anaerobic digestion process

Previous studies have shown that macro algae like *Ulva lactuca, Gracillaria vermiculophylla, Saccharina latissima* etc. can be anaerobically digested producing methane at yields ranging from 0.1-0.3 LCH₄/g volatile solids (VS) [43]. Methane yields of microalgae like *Spirulina platensis* (fresh water), and *Scenedesmus* spp. and *Chlorella* spp. (fresh water) ranged between 0.2 and 0.3 L CH₄/g VS [44, 45] when these were codigested with other feedstocks like dairy manure and waste paper sludge, whereas other microalgae like *Tetraselmis sp* (marine), *Chlorella vulgaris* (fresh water), *Scendesmus obliquess* (fresh water) and *Phaeodactylum tricornutum* (fresh water) produced an average methane yield ranging from 0.17 to 0.28 L CH₄/g VS [45-47] when digested as sole feedstock. Table 3 summarizes microalgae digestion studies reported in the literature. The Table also lists the methane yield of cellulose powder as a benchmark to compare the methane potentials of microalgae feedstocks. Depending on the type of microalgae, the methane potentials range from 5 to 78% of methane potential of cellulose. Choice of microalgae has an impact on the methane yield.

More recently when *Nannochloropsis oculata* was biogasified [48] in laboratory scale digesters at thermophilic temperature, the methane yield obtained was 0.20 L at STP/g VS. *N. oculata* was chosen because it can be grown easily in brackish or seawater, has a satisfactory growth rate and can tolerate a wide range of pH (7-10) and temperature (17 – 27º C). *N. oculata* is not rich in lipids but contains predominantly cellulose and other carbohydrates, which makes it a good feedstock for anaerobic digestion instead of biodiesel production. On a % (w/w dry matter) basis, the composition of *N. oculata* is: 7.8% carbohydrate, 35% protein and 18% lipid. Rest of the components are amino acids, fatty acids, omega-3, unsaturated alcohols, ascorbic acid [49]. About 88% of the carbohydrate is polysaccharide. Of the polysaccharides, 68.2% is glucose, and the rest are fucose, galactose, mannose, rhamnose, ribose and xylose.

Based on *N. oculata* growth observed in the pilot raceways and the methane yield from digestion of this alga, an analysis was carried out to estimate energy production and land requirements. Currently the algae harvesting rate from the raceways are 9.64 g ash free dry weight (afdw)/m^2/d. Note that afdw (ash free dry weight) is the same as volatile solids content. An often cited study for algae growth has yielded a much higher productivity of 50 g afdw/m^2/d for *Platyomonas sp* [50]. The algae biomass yield obtained in this study was only about 20% of the productivity potentially attainable. Optimization of growth conditions for *N. oculata* may improve its productivity. Using the methane yield value of 204 L/kg VS for anaerobic digestion of *N. oculata*, the annual energy output from a facility that grows the algae and subsequently digests it would be 27 MJ/m^2/year. The area occupied (or footprint) of the digester(s) would be far less than the land area required for growing the algae. If the methane produced from this facility is converted to electricity, the electrical energy output would be 2.25 kWH$_e$/m^2/year assuming that the efficiency of converting thermal energy to electrical energy is 30%. The household electrical energy and natural gas consumption in the US for the year 2010 was 11,496 kWH/year and 2070 m^3/year respectively. If the algae biogasification facility were to supply the entire electrical energy requirements for a household, the land area required would be 5108 m^2 (1.26 acres). If in addition, the facility were to supply the natural gas needs, then an additional 2900 m^2 (0.77 acres) would be needed. In other words ~2 acres of land could supply all the energy needs of a household in US. If the algae productivities were improved then land requirement could be further reduced. At 50 g afdw/m^2/d algae productivity, the land requirement would only be about 0.4 acres.

Despite useful methane production potential from biogasification and the ability to process dilute algal slurries in a digester, there are challenges to be overcome to commercialize this approach for producing bioenergy from microalgae. One bottleneck is that some feedstock characteristics can adversely affect anaerobic digestion. Unlike defined cultures used for production of biofuels like ethanol or butanol, the microbial consortia in an anaerobic digester is capable of secreting extracellular enzymes to hydrolyze and solubilize macromolecules like cellulose, hemicellulose, proteins and fats. This characteristic has enabled several terrestrial biomass feedstocks like sugarbeets, sugarbeet tailings, napier grass, sorghum and aquatic biomass like water hyacinth and giant kelp to be successfully digested using practical retention times. However, degradability of feedstocks containing high fraction of lignin (for example sugarcane bagasse, switchgrass, miscanthus and woody biomass like pine, eucalyptus) is poor in an anaerobic digester. The refractoriness of these feedstocks has been attributed to low moisture, crystalline nature of the cellulose, and complex association of the component carbohydrates within lignin [51]. As seen from Table 3, the digestibility of microalgae varies. Species with no cell wall or cell encapsulation composed of proteins like *Chlorella vulgaris* and *Phaeodactylum tricornutum*, has a higher yield of methane. *Dunaliella tertiolecta* has very low methane yield of 0.018 L/kg VS due to the presence of a cell wall consisting of cellulose fibers distributed within an organic matrix. So depending on the type of microalgae used it may be necessary to carry out some form of pretreatment of algae to improve methane yield and rate of methane production. The type of pretreatment may depend on algae type.

Strain	Source	Pretreatment	Digester operating conditions	Methane Yield L/kg VS	Reference
Chlorella vulgaris°	Fresh	None	No co-digestion Digestion at 30±5° C	0.22	[47]
Tetraselmis sp.	Marine	None	No co-digestion Digestion at 35°C	0.25	[46]
Scendesmus obliquus	Fresh	None	Hybrid flow through at 33±2°C and 54±2°C	0.17	[48]
Phaeodactylum tricornutum	Marine	None	Hybrid flow through at 33±2°C and 54±2°C	0.28	[48]
Dunaliella tertiolecta°	Marine	None	Serum bottle at 37°C	0.018	[47]

°Sample dried and then frozen at -24°C

Table 3. Summary of microalgae anaerobic digestion studies

6. Conclusion

Aqueous and marine biomass can be processed into a variety of sources of energy. Due to the extreme dilution in water, non-thermal processes such as anaerobic digestion, fermentation to bioalcohols, and lipid extraction are logical and useful methods to utilize key components of microorganisms to produce biofuels for the replacement or supplementing of traditional fossil fuels. However, thermal methods such as gasification of wet biomass may play a role in producing specialty fuels such as jet fuel that require a specific ratio of higher hydrocarbons that would prove otherwise difficult to manufacture, even given the requirement of intense drying.

In order for biofuels sourced from aqueous and marine biomass to secure a market share in the world, research and development needs to further nature's ability to produce higher concentrations of biomass with targeted characteristics and reduced footprints, while better utilizing available nutrients. This will allow for an ample supply of biomass to be produced without competition with the human food chain, that can be used renewably produce fuel that can power the world's mobile fleet.

Author details

Robert Diltz[1] and Pratap Pullammanappallil[2]

*Address all correspondence to: Robert.diltz@us.af.mil

1 Air Force Research Laboratory, Tyndall AFB, FL, USA

2 Department of Agricultural and Biological Engineering, University of Florida, Gainesville, FL, USA

References

[1] U.S. Energy Information Administration. International Energy Statistics. http://www.eia.gov/cfapps/ipdbproject/IEDIndex3.cfm?tid=5&pid=5&aid=2 (accessed August 6, 2012).

[2] Illman A, Scragg A, Shales S. Increase in Chlorella strains calorific values when grown in low nitrogen medium. *Enzyme and Microbial Technology* 2000; 27 631-635.

[3] Pan P, Hu C, Yang W, Li Y, Dong L, Zhu L, Tong D, Qing R, Fan Y. The direct pyrolysis and catalytic pyrolysis of *Nannochloropsis sp* residue for renewable bio-oils. *Bioresource Technology* 2010; 101 4593-4599.

[4] Collet C, Adler N, Schwitzguebel J, Peringer P. Hydrogen production by *Clostridium thermolacticum* during continuous fermentation of lactose. *International Journal of Hydrogen Energy* 2004; 29 1479-1485.

[5] Iakovou E, Karagiannidis A, Vlachos D, Toka A, Malamakis A. Waste biomass-to-energy supply chain management: A critical synthesis. *Waste Management* 2010; 30 1860-1870.

[6] Virmond E, Schacker R, Albrecht W, Althoff C, Souza M, Moreira R, Jose H. Organic solid waste originating from the meat processing industry as an alternative energy source. *Energy* 2010; 36 3897-3906.

[7] Liu A, Chen W, Song L. Identification of high-lipid producers for biodiesel production from forty-three green algal isolates in China. *Progress in Natural Science: Materials International* 2011; 21 269-276.

[8] Mujtaba G, Choi W, Lee C-G, Lee K. Lipid production by Chlorella vulgaris after a shift from nutrient-rich to nitrogen starvation conditions. *Bioresource Technology* 2012; doi: http://dx.doi.org/10.1016/j.biortech.2012.07.057.

[9] Griffiths M, Harrison T. Lipid productivity as a key characterisitic for choosing algal species for biodiesel production. *J Appl Phycol* 2009; 21 493-507.

[10] Araujo G, Matos L, Goncalves L, Fernandes F, Farias W. Bioprospecting for oil producing microalgal strains; Evaluation of oil and biomass production for tenn microalgal strains. *Bioresource Technology* 2011; 102 5248-5250.

[11] Borugadda V, Goud V. Biodiesel production from renewable feedstocks: Status and opportunities. *Renewable and Sustainable Energy Reviews* 2012; 4763-4784.

[12] Miron A, Garcia M, Camacho F, Grima E, Chisti Y. Growth and biochemical characterization of microalgal biomass produced in bubble column and airlift photobioreactors; studies in fed-batch culture. *Enzymy and Microbial Technology* 2002; 31 1015-1023.

[13] Khotimchenko S, Vaskovsky V, Titlyanova T. Fatty acids of marine algae from the Pacific Coast of North Carolina. *Botanica Marina* 2002; 45 17-22.

[14] Li X, Fan X, Han L, Lou Q. Fatty acids of some algae from the Bohai Sea. *Phytochemistry* 2002; 59 157-161.

[15] Gressler V, Yokoya N, Fujii M, Colepicolo P, Filho J, Torres R, Pinto E. Lipid, fatty acid, protein, amino acid and ash contents in four Brazilian red algae species. *Food Chemistry* 2010; 120 585-590.

[16] Courchesne N, Parisien A, Wang B, Lan C. Enhancement of lipid production using biochemical, genetic and transcription factor engineering approaches. *Journal of Biotechnology* 2009; 141 31-41.

[17] Lv J, Cheng L, Xu X, Zhang L, Chen H. Enhanced lipid production of *Chlorella vulgaris* by adjustment of cultivation conditions. *Bioresource Technology* 2010; 101 6797-6804.

[18] Zhao G, Y J, Jiang F, Zhang X, Tan T. The effect of different trophic modes on lipid accumulation of *Scenedismus quadricauda*. *Bioresource Technology* 2012; 114 466-471.

[19] Devi M, Subhash G, Mohan S. Heterotrophic cultivation of mixed microalgae for lipid accumulation and wastewater treatment during sequential growth and starvation phases: Effect of nutrient supplementation. *Renewable Energy* 2012; 43 276-283.

[20] Balat M, Balat M, Kirtay E. Main routes for the thermo-conversion of biomass into fuels and chemicals. Part 2: Gasification systems. *Energy Conversion and Management* 2009; 50 3158-3168.

[21] Wang L, Weller C, Jones D, Hanna M. Contemporary issues in thermal gasification of biomass and its application to electricity and fuel production. *Biomass and Bioenergy* 2008; 32 573-581.

[22] Akhtar J, Amin N. A review on process conditions for optimum bio-oil yield in hydroghermal liquefaction of biomass. *Renewable and Sustainable Energy Reviews* 2011; 15 (3) 1615-1624.

[23] Rustamov V, Abdullayev K,Samedov E. Biomass conversion to liquid fuel by two-stage thermochemical cycle. *Energy Conversion and Management* 1998; 39 (9) 869-875.

[24] Xu C, Etcheverry T. Hydro-liquefaction of woody biomass in sub- and super-critical ethanol with iron-based catalysts. *Fuel* 2008; 87 (3) 335-345.

[25] Sun P, Heng M, Sun S, Chen J. Direct liquefaction of paulownia in hot compressed water: Influence of catalysts. *Energy* 2010; 35 5421-5429.

[26] Balat M, Balat M, Kirtay E. Main routes for the thermo-conversion of biomass into fuels and chemicals. Part 1: Pyrolysis systems. *Energy Conversion and Management* 2009; 50 3147-3157.

[27] Onay O, Kockar O. Slow, fast and flash pyrolysis of rapeseed. *Renewable Energy* 2003; 28 2417-2433.

[28] Hajaligol M, Waymack B, Kellogg D. Low temperature formation of aromatic hydrocarbon from pyrolysis of cellulosic materials. *Fuel* 2001; 80 1799-1807.

[29] M.F. Demirbas, M. Balat, and H. Balat, *Energy Conversion and Management* 50, 1746 (2009).

[30] Smets K, Adriaensens P, Reggers G, Schreurs S, Carleer R, Yperman J. Flash pyrolysis of rapeseed cake: Influence of temperature on the yield and the characteristics of the pyrolysis liquid. *Journal of Analytical and Applied Pyrolysis* 2011; 90 118-125.

[31] Fowler P, Krajacic G, Loncar D, Duic N. Modeling the energy potential of biomass-H₂RES. *Hydrogen Energy* 2009; 34 7027-7040.

[32] Kelly-Yong T, Lee K, Mohamed A, Bhatia S. Potential of hydrogen from oil palm biomass as a source of renewable energy worldwide. *Energy Policy* 2007; 35 5692-5701.

[33] Baier K, Dühring I, Enke H, Gründel M, Lockau W, Smith C, Woods P, Ziegler K, Oesterhelt C, Coleman J,Kramer D. Genetically modified photoautotrophic ethanol producing host cells, methods for producing the host cells, constructs for the transformation of the host cells, method for testing a photoautotrophic strain for desired growth property and method for producing ethanol using the host cells. International Application No. PCT/EP 2009/000892 (2009).

[34] Woods, P. Algenol Biofuels' Direct to Ethanol Technology AIChE, Nashville, TN (2009).

[35] Woods R, Legere E, Moll B, Unamunzanga C, Mantecon E. Closed photobioreactor system for continued daily in situ production, separation, collection and removal of

ethanol from genetically enhanced photosynthetic organisms. US Patent 7,682,821 (2007).

[36] Luo D, Hu Z, Choi D, Thomas V, Realff M, Chance R. Life cycle energy and greenhouse gas emissions for an ethanol production process based on blue-green algae. *Environ. Sci. Technol.* 2010; 44 8670-8677.

[37] Sialve B, Bernet N, Bernard O. Anaerobic digestion of microalgae as a necessary step to make microalgal biodiesel sustainable. *Biotechnology Advances* 2009; 27 409–416.

[38] Singhal V, Rai J. Biogas production from water hyacinth and channel grass used for phytoremediation of industrial effluents. *Bioresource Technology* 2003; 86 221–225.

[39] Brown M. The amino-acid and sugar composition of 16 species of microalgae used in mariculture. *Journal of Experimental Marine Biology and Ecology* 1991; 145 79–99.

[40] Elser J, Fagan W, Denno R, Dobberfuhl D, Folarin A, Huberty A, Interlandi S, Kilham S, McCauley E, Schulz K, Siemann E, Sterner R. Nutritional constraints in terrestrial and freshwater food webs. *Nature* 2000; 408 578–580.

[41] Droop M. 25 Years of Algal Growth Kinetics A Personal View. *Botanica Marina* 1983; 26 99–112.

[42] Grobbelaar J. Algal Nutrition - Mineral Nutrition. *Handbook of Microalgal Culture: Biotechnology and Applied Phycology* 2007; 95–115.

[43] Nielsen H, Heiske S. Anaerobic digestion of macroalgae: methane potentials, pretreatment, inhibition and co-digestion. *Water Science & Technology* 2011; 64 1723-1729.

[44] Sánchez Hernández E, Travieso Córdoba L. Anaerobic digestion of *Chlorella vulgaris* for energy production. *Resources, Conservation and Recycling* 1993; 9 127–132.

[45] Marzano C, Legros A, Naveau H, Nyns E. Biomethanation of the Marine Algae *Tetraselmis*. *International Journal Of Solar Energy* 1982; 1 263–272.

[46] Lakaniemi A, Hulatt C, Thomas D, Tuovinen O, Puhakka J. Biogenic hydrogen and methane production from *Chlorella vulgaris* and *Dunaliella tertiolecta* biomass. *Biotechnology for biofuels* 2011; 34 (4).

[47] Zamalloa C, Boon N, Verstraete W. Anaerobic digestibility of *Scenedesmus obliquus* and *Phaeodactylum tricornutum* under mesophilic and thermophilic conditions. *Applied Energy* 2012; 92 733-738.

[48] Buxy S, Diltz R, Pullammanappallil P. Biogasification of marine algae *Nannochloropsis oculata*. Accepted for publication in Proceedings of Material Challenges in Alternative and Renewable Energy. Trans of American Ceramic Society (2012).

[49] Foree E, McCarty P. Anaerobic decomposition of algae. *Environ. Sci. Technol.* 1970; 4 842–849.

[50] Sheehan J, Dunahay T, Benemann J, Roessle P. A Look Back at the U.S. Department of Energy's Aquatic Species Program—Biodiesel from Algae. National Renewable Energy Laboratory report, July 1998.

[51] Chynoweth D, Jerger D. Anaerobic digestion of woody biomass. *Developments in Industrial Microbiology* 1985; 26 235-246.

Chapter 2

Biofuel: Sources, Extraction and Determination

Emad A. Shalaby

Additional information is available at the end of the chapter

1. Introduction

Biofuel is a type of fuel whose energy is derived from biological carbon fixation. Biofuels include fuels derived from biomass conversion (Figure 1, JICA, Okinawa, Japan), as well as solid biomass, liquid fuels and various biogases. Although fossil fuels have their origin in ancient carbon fixation, they are not considered biofuels by the generally accepted definition because they contain carbon that has been "out" of the carbon cycle for a very long time. Biofuels are gaining increased public and scientific attention, driven by factors such as oil price hikes, the need for increased energy security, concern over greenhouse gas emissions from fossil fuels, and support from government subsidies. Biofuel is considered carbon neutral, as the biomass absorbs roughly the same amount of carbon dioxide during growth, as when burnt. The chemical composition of different kinds of biomass was shown in Table 1.

Biodiesel as one from important biofuel types is made from vegetable oils and animal fats. Biodiesel can be used as a fuel for vehicles in its pure form, but it is usually used as a diesel additive to reduce levels of particulates, carbon monoxide, and hydrocarbons from diesel-powered vehicles. Biodiesel is produced from oils or fats using transesterification and is the most common biofuel in Europe.

Bioethanol is an alcohol made by fermentation, mostly from carbohydrates produced in sugar or starch crops such as corn or sugarcane. Cellulosic biomass, derived from non-food sources such as trees and grasses, is also being developed as a feedstock for ethanol production. Ethanol can be used as a fuel for vehicles in its pure form, but it is usually used as a gasoline additive to increase octane and improve vehicle emissions. Bioethanol is widely used in the USA and in Brazil. Current plant design does not provide for converting the lignin portion of plant raw materials to fuel components by fermentation.

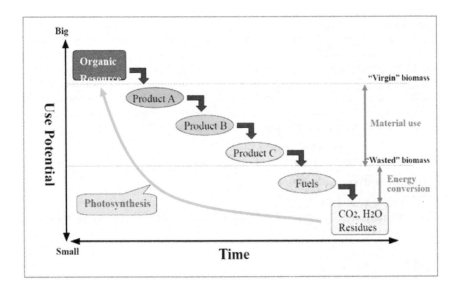

Figure 1. Cascade use of biomass

Kinds of bimass			Characteristics								
			Moisture content (%)	Solid substance(%)				Minerals(dry matter%)			
				Organic matter			Minerals	Carbon (T-C)	Nitrogen (T-N)	Phosphrus (T-P)	Potasium(T-K)
				Bacteria decomposi-tion	Hard decomposi-tion	Total					
Livestock waste	Beef cow dung		78.0	5.3	12.3	17.6	4.4	35.1	2.5	1.0	1.5
	Dairy cow	Dung	85.0	3.6	8.4	12.0	3.0	35.1	2.5	1.0	1.5
		Mixed	90.0	2.4	5.6	8.0	2.0	35.1	2.5	1.0	1.5
	Pig	Dung	72.0	10.5	10.5	21.0	7.0	35.1	3.5	2.5	1.5
		Mixed	90.0	3.8	3.8	7.6	2.5	35.1	3.5	2.5	1.5
	Poultry		70.0	15.8	5.2	21.0	9.0	35.1	5.5	3.0	3.0
Food residue	Tofu grounds		80.0	17.4	1.6	19.0	1.0	44.2	4.5	0.5	1.0
Garbage	Garbage		80.0	14.2	2.5	16.7	3.3	44.2	2.5	2.4	3.4
	Rubbish		50.0	29.4	12.6	42.0	8.0	55.0	1.0	—	—
	Used cooking oil		—	—	—	—	—	71.4	—	—	—
Slurry	Human waste		97.8	0.7	1.0	1.7	0.5	38.4	4.5	2.1	0.2
	Septic tank		99.0	0.3	0.5	0.8	0.2	38.4	4.5	1.1	0.1
	Sewage		78.0	5.8	8.6	14.4	7.6	38.4	4.5	2.3	0.3
Crop-wood residue	Wood residue		35.0	12.8	51.2	64.0	1.0	52.0	0.4	0.02	0.1
	Crop residue		12.0	32.0	38.7	70.7	17.3	40.9	0.6	0.09	0.4

Table 1. Average properties of biomass

In 2010 worldwide biofuel production reached 105 billion liters (28 billion gallons US), up 17% from 2009, and biofuels provided 2.7% of the world's fuels for road transport, a contribution largely made up of ethanol and biodiesel. Global ethanol fuel production reached 86 billion liters (23 billion gallons US) in 2010, with the United States and Brazil as the world's top producers, accounting together for 90% of global production. The world's largest biodiesel producer is the European Union, accounting for 53% of all biodiesel production in 2010. As of 2011, mandates for blending biofuels exist in 31 countries at the national level and in 29 states/provinces. According to the International Energy Agency, biofuels have the potential to meet more than a quarter of world demand for transportation fuels by 2050.

2. Different sources of biofuel

Here are 4 biofuel sources, with some of their application in developmental stages, some actually implemented:

2.1. Algae

Algae come from stagnant ponds in the natural world, and more recently in algae farms, which produce the plant for the specific purpose of creating biofuel. *Advantage of algae focude on the followings:* No CO_2 back into the air, self-generating biomass, Algae can produce up to 300 times more oil per acre than conventional crops. Among other uses, algae have been used experimentally as a new form of green jet fuel designed for commercial travel. At the moment, the upfront costs of producing biofuel from algae on a mass scale are in process, but are not yet commercially viable (Figure 2)

2.2. Carbohydrate (sugars) rich biomaterial

It comes from the fermentation of starches derived from agricultural products like corn, sugar cane, wheat, beets, and other existing food crops, or from inedible cellulose from the same. Produced from existing crops, can be used in an existing gasoline engine, making it a logical transition from petroleum. It used in Auto industry, heating buildings ("flueless fireplaces") At present, the transportation costs required to transport grains from harvesting to processing, and then out to vendors results in a very small net gain in the sustainability stakes.

2.3. Oils rich biomaterial

It comes from existing food crops like rapeseed (aka Canola), sunflower, corn, and others, after it has been used for other purposes, i.e food preparation ("waste vegetable oil", or WVO), or even in first use form ("straight vegetable oil", or SVO). Not susceptible to microbial degradation, high availability, re-used material. It is used in the creation of biodiesel fuel for automobiles, home heating, and experimentally as a pure fuel itself. At present, WVO or SVO is not recognized as a mainstream fuel for automobiles. Also, WVO and SVO are susceptible to low temperatures, making them unusable in colder climates.

TRENDS in Ecology & Evolution

Figure 2. Dense algal growth in four pilot-scale tank bioreactors fed by treated wastewater from the Lawrence, Kansas (USA) wastewater treatment plant (photo by B. Sturm). Each fiberglass bioreactor has an operating volume of ten cubic meters of water, and is operated as an air-mixed, flow-through vessel. Nutrientrich wastewater inflows are pumped in through the clear plastic hose (blue clamp), and water outflow occurs through the white plastic pipe shown at the waterline. These bioreactors are intended to be operated year-round, as the temperature of the inflowing wastewater is consistently ca. 10 - 8°C.

2.4. Agriculture wastes (organic and inorganic sources)

It comes from agricultural waste which is concentrated into charcoal-like biomass by heating it. Very little processing required, low-tech, naturally holds CO_2 rather than releasing it into the air. Primarily, biochar has been used as a means to enrich soil by keeping CO_2 in it, and not into the air. As fuel, the off-gasses have been used in home heating. There is controversy surrounding the amount of acreage it would take to make fuel production based on biochar viable on a meaningful scale. Furthermore, use of agriculture wastes which rich with inorganic elements (NPK----) as compost (fertilizer) in agriculture.

3. Comparison between different extraction methods of bio-diesel, bio-ethanol, biogas (bio-methane)

3.1. Biodiesel

3.1.1. Biodiesel extraction

Biodiesel is a clean-burning diesel fuel produced from vegetable oils, animal fats, or grease. Its chemical structure is that of fatty acid alkyl esters (FAAE). Biodiesel as a fuel gives much lower toxic air emissions than fossil diesel. In addition, it gives cleaner burning and has less sulfur content, and thus reducing emissions. Because of its origin from renewable resources, it is more likely that it competes with petroleum products in the future. To use biodiesel as a fuel, it should be mixedwith petroleum diesel fuel to create a biodiesel-blended fuel. Biodiesel refers to the pure fuel before blending. Commercially, biodiesel is produced by transesterification of triglycerides which are the main ingredients of biological origin oils in the presence of an alcohol (e.g. methanol, ethanol) and a catalyst (e.g. alkali, acid, enzyme) with glycerine as a major by-product [Ma and Hanna, 1999 ; Dube et al., 2007]. After the reaction, the glycerine is separated by settling or centrifuging and the layer obtained is purified prior to using it for its traditional applications (pharmaceutical, cosmetics and food industries) or for the recently developed applications (animal feed, carbon feedstock in fermentations, polymers, surfactants, intermediates and lubricants) [Vicente et al., 2007].

However, one of the most serious obstacles to use biodiesel as an alternative fuel is the complicated and costly purification processes involved in its production. Therefore, biodiesel must be purified before being used as a fuel in order to fulfil the EN 14214 and ASTM D6751 standard specifications listed in Table 2; otherwise the methyl esters formed cannot be classified as biodiesel. Removing glycerine from biodiesel is important since the glycerine content is one of the most significant precursors for the biodiesel quality. Biodiesel content of glycerine can be in the form of free glycerine or bound glycerine in the form of glycerides. In this work we refer to the total glycerine, which is the sum of free glycerine and bound glycerine. Severe consequences may result due to the high content of free and total glycerine, such as buildup in fuel tanks, clogged fuel systems, injector fouling and valve deposits (Hayyan et al., 2010).

3.1.2. Biodiesel extraction methods:

3.1.2.1. One step transesterification

For the synthesis of biodiesel, the following materials were used: oil sample (FFM Sdn Bhd), methanol (Merck 99%), and potassium hydroxide (KOH) as a catalyst (HMGM Chemicals >98%). Methanol and potassium hydroxide were pre-mixed to prepare potassium methoxide, and then added to oil in the reactor with a mixing speed of 400 rpm for 2 h at 50 °C. The molar ratio of oil to methanol was 1:10. Finally, the mixture was left overnight to settle forming two layers, namely: biodiesel phase (upper layer) and the glycerin-rich phase (Figure 3).

Property	EN 14214		ASTM D 6751	
	Test method	Limits	Test method	Limits
Ester content	EN 14103	96.5% (mol mol^{-1}) min	-	-
Linolenic acid content	EN 14103	12.0% (mol mol^{-1}) max	-	-
Content of FAME[a] with ≥4 double bonds	-	1.0% (mol mol^{-1}) max	-	-
MAG[b] content	EN 14105	0.80% (mol mol^{-1}) max	-	-
DAG[c] content	EN 14105	0.20% (mol mol^{-1}) max	-	-
TAG[d] content	EN 14105	0.20% (mol mol^{-1}) max	-	-
Free glycerine	EN 14105	0.02% (mol mol^{-1}) max	ASTM D 6584	0.020% (w/w) max
Total glycerine	EN 14105	0.25% (mol mol^{-1}) max	ASTM D 6584	0.240% (w/w) max
Water and sediment or water content	EN ISO 12937	500 mg kg^{-1} max	ASTM D 2709	0.050% (v/v) max
Methanol content	EN 14110	0.20% (mol mol^{-1}) max	-	-
(Na + K) content	EN 14108	5.0 mg kg^{-1} max	UOP 391	5.0 mg kg^{-1} max
(Ca + Mg) content	prEN 14538	5.0 mg kg^{-1} max	-	-
P content	EN 14107	10.0 mg kg^{-1} max	ASTM D 4951	0.001% (w/w) max
Oxidative stability (110 °C)	EN 14112	6 h min	-	-
Density (15 °C)	EN ISO 3675	860–900 kgm^{-3}	-	-
Kinematic viscosity or viscosity (40 °C)	EN ISO 3104	3.5–5.0 mm^2 s^{-1}	ASTM D 445	1.9–6.0 mm^2 s^{-1}
Flash point	EN ISO 3679	120 °C min	ASTM D 93	130 °C min
Cloud point	-	-	ASTM D 2500	Not specified
Sulfur content	EN ISO 20864	10.0 mg kg^{-1} max	ASTM D 5453	0.05% (w/w) max
Carbon residue	EN ISO 10370	0.30% (mol mol^{-1}) max	ASTM D 4530	0.050% (w/w) max
Cetane number	EN ISO 5165	51 min	ASTM D 613	47 min
Sulphated ash	ISO 3987	0.02% (mol mol^{-1}) max	ASTM D 874	0.020% (w/w) max
Total contamination	EN 12662	24 mg kg^{-1} max	-	-
Copper strip corrosion (3 h, 50 °C)	EN ISO 2160	1 (degree of corrosion	ASTM D 130	No. 3 max
Acid number or acid value	EN 14104	0.50 mg KOH g^{-1} max	ASTM D 664	0.50 mg KOH g^{-1} max
Iodine value	EN 14111	120 g I2·100 g^{-1} max	-	-
Distillation temperature (90% recovered)	-	-	ASTM D 1160	360 °C max

a FAME = fatty acid methyl esters.
b MAG = monoacylglycerines.
c DAG = diacylglycerines.
d TAG = triacylglycerine.

Table 2. Biodiesel specifications according to EN 14214, and ASTM D6751 standards.

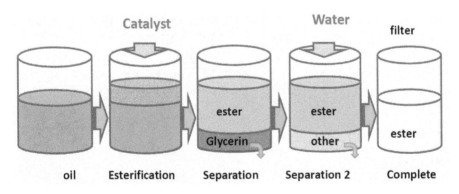

Figure 3. The biodiesel extraction process (steps).

3.1.2.2. Second step transterification

The production methodology followed in this study was according to Tomosevic and Si-ler-Marinkovic [2003] with some modification, where the alkali-catalyzed transesterifica-tion was applied. Basically, methanol was the alcohol of choice and KOH was used as the catalyst. Potassium methoxide solution (PMS) was prepared freshly by mixing a pre-determined amount of methanol ($\approx$ 12 wt % of oil) with KOH ($\approx$ 1.0 wt % of oil) in a container until all the catalyst dissolved. The PMS was then added to 200 g of oil and stirred vigorously for 30 min at 30°C. Then after, the mixture was carefully transferred to a separating funnel and allowed to stand for 4 h. the lower layer (glycerol, methanol and most of the catalysts) was drained out. The upper layer (methyl esters MEs, some methanol and traces of the catalyst) was transferred into another flask containing freshly prepared PMS mixed at 60 rpm under reflux at 60°C for 30 min. afterwards; the mixture was carefully transferred to a separating funnel and allowed to stand there over night. The glycerol was removed by gravity settling, whereas the obtained crude esters layer was transferred into water bath to remove excess methanol at 65°C and 20 kPa. The obtained crude methyl esters were then cleaned thoroughly by washing with warm (50°C) deionized water, dried over anhydrous Na_2SO_4, weighted and applied for further analysis (Shalaby and Nour, 2012; Shalaby, 2011).

3.1.2.3. Qualitative analysis of glycerol

The Borax/phth test is special test for detection on the compound contain two neighboring hydroxyl group as in glycerol organic compound as the following:

1 ml glycerol layer mix with 1 ml of Borax/phth (red color) if the red color disappear in cold and appearing after heating (direct) this positive control.

3.1.2.4. Fourier transforms infrared spectroscopy (FTIR) analysis

FTIR analysis was performed using instrument, Perkin Elmer, model spectrum one, for detection of transesterification efficiency of oil by determination of the active groups produced from these process.

The results obtained by Shalaby and Nour (2012) found that, two step transterification of oil led to 100 % disappearance of hydroxyl group but this was less than 100 % in case of one step transterification as shown in Figure (4).

3.2. Bioethanol

3.2.1. Bioethanol extraction

Bioethanol is one of the most important renewable fuels due to the economic and environmental benefits of its use. The use of bioethanol as an alternative motor fuel has been steadily increasing around the world for the number of reasons. 1) Fossil fuel resources are declining, but biomass has been recognized as a major reasons World renewable energy source. 2) Greenhouse gas emissions is one of the most important challenges in this century because of fossil fuel consumption, biofuels can be a good solution for this problem. 3) Price of petroleum in global market has raising trend. 4) Petroleum reserves are limited and it is monopoly of some oil-importing countries and rest of the world depends on them. 5) Also known petroleum reserves are estimated to be depleted in less than 50 years at the present rate of consumption. At present, in compare to fossil fuels, bioethanol is not produced economically, but according to scientific predictions, it will be economical about 2030.

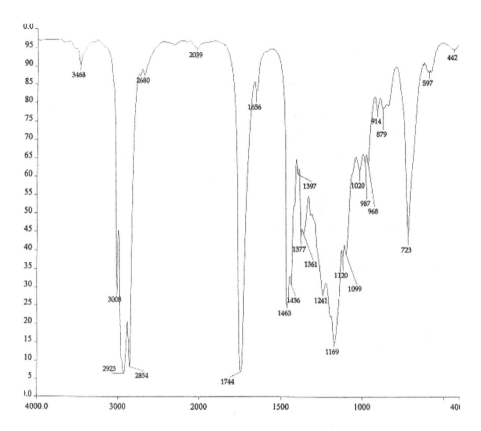

Figure 4. The IR spectrum of oil after two step transterification (produced biodiesel) process

Biomass commonly gathers from agricultural, industrial and urban residues. The wastes used for bioethanol production are classified in three groups according to pretreatment process in sugary, starchy and lignocellulosic biomasses. Lignocellulosic biomass, including forestry residue, agricultural residue, yard waste, wood products, animal and human wastes, etc., is a renewable resource that stores energy from sunlight in its chemical bonds. Lignocellulosic biomass typically contains 50%-80% (dry basis) carbohydrates that are polymers of 5C and 6C sugar units. Lignocellulosic biomasses such as waste wood are the most promising feedstock for producing bioethanol.

Bioconversion of lignocellulosic biomass to ethanol is significantly hindered by the structural and chemical complexity of biomass, which makes these materials a challenge to be used as feedstock for cellulosic ethanol production. Cellulose and hemicellulose, when hydrolyzed into their component sugars, can be converted into ethanol through well-established fermentation technologies. However, sugars necessary for fermentation are trapped inside the crosslinking structure of the lignocellulose.

Conventional methods for bioethanol production from lignocellulosic biomasses take three steps: *pretreatment* (commonly acid or enzyme hydrolyses), *fermentation, distillation*. Pretreatment is the chemical reaction that converts the complex polysaccharides to simple sugar. pretreatment of biomass is always necessary to remove and/or modify the surrounding matrix of lignin and hemicellulose prior to the enzymatic hydrolysis of the polysaccharides (cellulose and hemicellulose) in the biomass. Pretreatment refers to a process that converts lignocellulosic biomass from its native form. In general, pretreatment methods can be classified into three categories, including physical, chemical, and biological pretreatment. In this step, biomass structure is broken to fermentable sugars. This project focused on chemically and biologically pretreatment. For example: this project shows the effect of sulfuric acid, hydrochloric acid and acetic acid with different concentration by different conditions also shows the effect of cellulase enzyme by different techniques. Then fermentation step in which there are a series of chemical or enzymatic reactions that converted sugar into ethanol. The fermentation reaction is caused by yeast or bacteria, which feed on the sugar such as *Saccharomyces cerevisae*. After that, distillation step in which the pure ethanol is separated from the mixture using distiller which boil the mixture by heater and evaporate the mixture to be condensate at the top of the apparatus to produce the ethanol from joined tube.

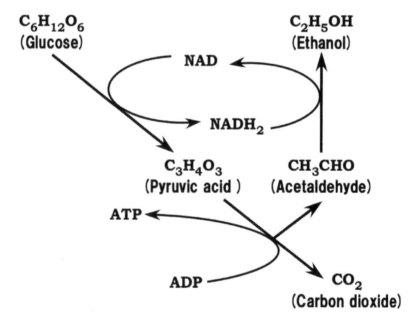

Figure 5. Ethanolic fermentation metabolism chart

The way to manufacture bioethanol is basically the same as that of liquor. Generally, saccharinity material such as sugar and starchy material such as rice and corn are saccharified (Figure 5-7), fermented and distilled till absolute ethanol whose alcoholicity is over 99.5%. It is technically possible to manufacture ethanol from cellulosic material such as rice straw or wood remains.

3.2.2. How to produce bio-ethanol:

• Materials

Sugarcane stems 5kg

Dry yeast, 15g

• Items

Brix meter, 5L flask, Dimroth condenser, Liebig condenser, Stick, Beaker

Cloth filter

1. Fermentation method

2. Mill juice out of Sugarcane stems. (about 3L of juice)

3. The juice is filtered out impurities.

4. Measurement Brix of juice.

5. Dry yeast is added to juice, the rate of 6g/L.

6. It keeps in the flask which sealed except the vent.

7. A cover is opened one day and once, then juice and dry yeast mixes so that air may enter with stick.

8. It continues until Brix becomes fixed.

9. Distillation method (Fig. 8)

10. Fermented juice is filtered out sediment.

11. It heats to boiling point in distiller.

12. Dimroth condenser is kept warm (about 70 degree) with hot water which is made to circulate by a pump.

13. Allihn condenser cools with tap water (about 20 degree).

14. Bio-ethanol which falls from the point of a allihn condenser is caught with beaker on ice.

3.2.3. Qualitative analysis for ethanol

Iodoform test on cold is special test for ethanol as the following: I ml ethanol layer mix with iodide and sodium hydroxide after that, the presence of yellow crystal and iodoform odor produced, this meaning presence of ethanol.

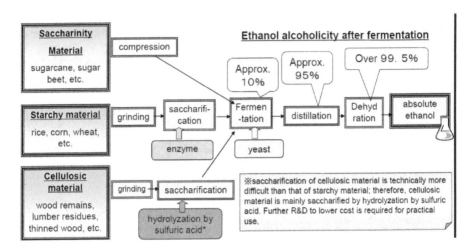

Figure 6. Production of absolute ethanol from Saccharinity, Starch and Cellulosic materials

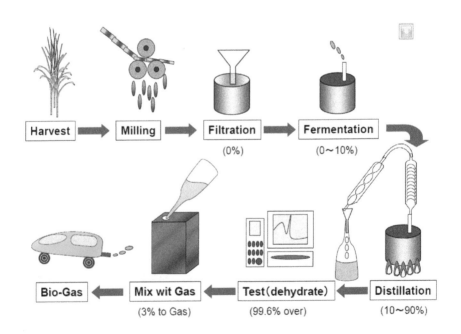

Figure 7. The main steps of bioethanol production from Starchy and cellulosic materials (Masami YASUNAKA / JIRCAS)

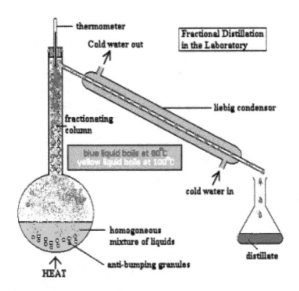

Figure 8. The distillation process for ethanol production.

3.2.4. Quantitative ethanol determination

3.2.4.1. Direct injected GC method

Beverage sample solution (0.5 mL) was dispensed into an l-mL caped sample vial, and then 5 mL of 1% internal standard solution (equivalent to 50 mg) was added. After mixing, 0.1 µL of the sample solution was injected directly into a GC or GC/MS (Figure 9) with syringe (Anonymous. 1992; Collins et al., 1997).

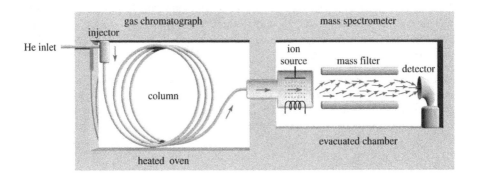

Figure 9. The GC/MS used for determination of ethanol.

3.2.4.2. Dichromate oxidation method

Beverage sample solution (1~5 mL) was steam distillated to obtain alcoholic eluate (> 50 mL), and then oxidized with acidified dichromate. The excessive potassium dichromate was then titrated with ferric oxide. The ethanol content in beverage sample could be obtained by calculating the volume difference of potassium dichromate consumption between sample solution and control solution (Anonymous. 1992; Collins et al., 1997).

3.2.4.3. Distillation-hydrometric method

Alcoholic volatile compounds in beverage samples were separated by distillation, and the gravity of the distillate was measured by hydrometer. The ethanol content was then converted (Anonymous. 1992; Collins et al., 1997).

3.3. Biogas (bio-methane) extraction

Methane fermentation is a versatile biotechnology capable of converting almost all types of polymeric materials to methane and carbon dioxide under anaerobic conditions. This is achieved as a result of the consecutive biochemical breakdown of polymers to methane and carbon dioxide in an environment in which varieties of microorganisms which include fermentative microbes (acidogens); hydrogen-producing, acetate-forming microbes (acetogens); and methane-producing microbes (methanogens) harmoniously grow and produce reduced end-products (Fig. 10-11). Anaerobes play important roles in establishing a stable environment at various stages of methane fermentation.

Methane fermentation offers an effective means of pollution reduction, superior to that achieved via conventional aerobic processes. Although practiced for decades, interest in anaerobic fermentation has only recently focused on its use in the economic recovery of fuel gas from industrial and agricultural surpluses.

The biochemistry and microbiology of the anaerobic breakdown of polymeric materials to methane and the roles of the various microorganisms involved are discussed here. Recent progress in the molecular biology of methanogens is reviewed, new digesters are described and improvements in the operation of various types of bioreactors are also discussed.

Methane fermentation is the consequence of a series of metabolic interactions among various groups of microorganisms. A description of microorganisms involved in methane fermentation, based on an analysis of bacteria isolated from sewage sludge digesters and from the rumen of some animals,. The first group of microorganisms secretes enzymes which hydrolyze polymeric materials to monomers such as glucose and amino acids, which are subsequently converted to higher volatile fatty acids, H_2 and acetic acid (Fig. 10). In the second stage, hydrogen-producing acetogenic bacteria convert the higher volatile fatty acids *e.g.*, propionic and butyric acids, produced, to H_2, CO_2, and acetic acid. Finally, the third group, methanogenic bacteria convert H_2, CO_2, and acetate, to CH_4 and CO_2 (Nagai et al., 1986).

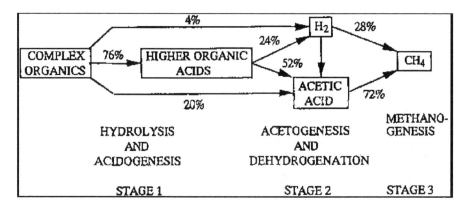

Figure 10. The main steps for production of methane gas

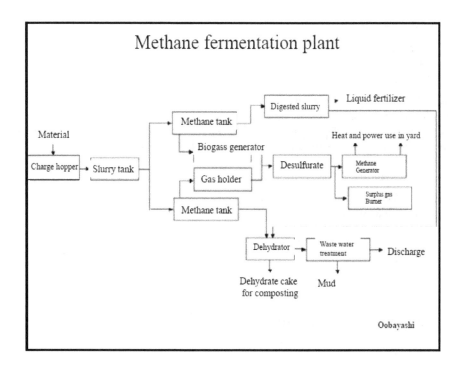

Figure 11. The principles methods for biomethane production

3.4. Determination of methane concentration

Methane will be measured on the gas chromatogram (Figure 9)using a FID (flame ionization) detector.

Note, unless you want smelly hands, it is recommended that you wear gloves. A lab coat is recommended for similar reasons.

- Using a 20 ml syringe connected to a 2-way stopcock, collect a little more than 5 ml of water from a port on your Winogradsky column.

- With the syringe pointing up, remove any air (tapping the sides of the syringe) and expel any extra water so that the final liquid volume in the syringe is 5 ml. Do this over a sink.

- Now, draw in 15 ml of air into the syringe so that the total air+water volume in the syringe is 20 ml. Close the stopcock.

- Shake the syringe to equilibrate the methane between the air and water.

- With the syringe pointing down, eject all the water from the syringe into the sink and close the stopcock. Try to get all the water out, but leave at least 10 ml of gas in the syringe

- We will now move to the GC lab in Starr 332 to measure methane.

- Repeat the above procedure for each of the ports on your Winogradsky column.

3.5. Calculations

To assist in plotting up results, measure the distance from the top of the sediment-water interface to each of the ports on the Winogradsky column, with distance to the ports in the sediment as positive and those in the water column negative. Also, measure the distance from the sediment-water interface to the surface of the water and the bottom of the sediments.

3.6. Methane concentration calculation

- From the standards, determine the concentration of methane in ppmv. Use the ideal gas law to determine the number of moles of methane in the 15 ml gas volume:

$$n = \frac{PV}{RT} = \frac{\frac{ppm}{10^6} \cdot \frac{15}{1000}}{(0.08205)(293)} \tag{1}$$

4. Physico-chemical parameters of extracted biofuel

4.1. Biodiesel

Most of the physical and chemical properties of the obtained methyl esters were determined by methods listed in JUS EN 14214:2004 standard [JUS EN 14214:2004] equivalent to EN 14214:2003, which defines requirements and test methods for fatty acid methyl esters (FAME) to be used in diesel engine. It must be emphasized that the characterization of crude methyl esters (i.e. those obtained before the purification) was not performed as it is well known fact that such raw products represent mixtures that were not in compliance with the strict restrictions for alternative diesel fuels, as it contains glycerol, alcohol, catalyst, mono- and diglycerides besides fatty acid esters. Measurements of the density at 15 _C by hydrometer method and of the kinematic viscosity at 40 _C were carried out according to JUS EN ISO 3675:1988 and JUS ISO 3104:2003, respectively. The acid value (Av) was determined by titration in accordance to EN 14104:2003; the iodine value was obtained by Hannus method (EN 14111:2003) this property has been also previously used for the biodiesel characterization [Karaosmanog et al., 1996; Šiler-Marinkovic et al., 1998]. The method for the cetane index (CI) estimation based on the saponification (Sv) and iodine (Iv) values was previously described [Krisnangkura, 1986] as simpler and more convenient than experimental procedure for the cetane number determination utilizing a cetane engine (EN ISO 5165:1998). The Krisnangkura's equation [Krisnangkura, 1986] used for CI calculation was as follows: CI = 46.3 + 5458/Sv_0.225 Iv. The cloud polint of MEs was determined according to ASTM D-2500 and Total sulfur content according to ASTM D-4294, Copper strip corrosion at 100 C according to ASTM D-130.The methyl ester composition was obtained by gas chromatograph equipped with DB-WAX 52 column (Supelco) and flame ionization detector. All the properties of frying oils as example were analyzed in two replicates and the final results given below were obtained as the average values (Table 3).

4.1.1. Density at 15 °C

It is known that biodiesel density mainly depends on its methyl esters content and the remained quantity of methanol (up to 0.2% m/m according to JUS EN 14214 [JUS EN 14214:2004]); hence this property is influenced primarily by the choice of vegetable oil [Mittelbach, 1996], and in some extent by the applied purification steps. the mean density value of produced biodiesel was 0.90 g/cm3, while this value was more than Egyptian diesel (0.82-0.87g/cm3). but met the density value specified by JUS EN 14214 [JUS EN 14214:2004] to be in the range 0.860–0.900 g/cm3 at 15 °C. This property is important mainly in airless combustion systems because it influences the efficiency of atomization of the fuel [Felizardo et al., 2006].

4.1.2. Kinematic viscosity at 40 °C

Even more than density, kinematic viscosity at 40 °C is an important property regarding fuel atomization and distribution. With regard to the kinematic viscosities that were in the range from 32.20 to 48.47 mm2/s, the feedstocks differed among themselves significantly. The viscosities of MEs were much lower than their respective oils (about 10 times) and they met the required values that must be between 3.5 and 5.0 mm2/s [JUS EN 14214:2004]. Comparing our MEs, the increase of the viscosities was observed more than Egyptian diesel, EN14214 and D-6751 (14.3, 7, 5 and 6 respectively) as shown in Table (3). However, the kinematic viscosity at 100 °C of MEs produced from frying oil was met the viscosity range of Egyptian diesel, EN14214 and D-6751 (4.3, 7, 5 and 6 respectively). Predojevic (2008).

4.1.3. Acid value

The acid value measures the content of free acids in the sample, which have influence on fuel aging. It is measured in terms of the quantity of KOH required to neutralize sample. The base catalyzed reaction is reported to be very sensitive to the content of free fatty acids, which should not exceed a certain limit recommended to avoid deactivation of catalyst, formation of soaps and emulsion [Sharma et al., 2008, Meher et al., 2004]. The feedstock acid values obtained in this study differed significantly ranging from 1.86 to 3.31 mg KOH/g oil. Thus, in the light of the previous discussion on the requirements for the feedstock acid values, it could be concluded that frying oil had the values above the recommended 2 mg KOH/g. However, these values did not turn out to be limiting for the efficiency of the applied two-stage process, as it will be discussed along to the obtained product yields and purity later on. Acid values of MEs were less than 0.5 mg KOH/g specified as the maximum value according to JUS EN14214 (Table 4), Sharma et al. (2008) reviewed the literature and found that acid value of the feedstock for alkaline transesterification has to be reduced to less than 2 mg KOH/g (i.e. 1%), while only few examples of transesterification with feedstock acid value of up to 4.0 mg KOH/g (i.e. 2%) were found. They also reported that when waste cooking oil is used as feedstock, the limit of free fatty acids is a bit relaxed and the value a little beyond 1% (i.e. 2 mg KOH/g) did not have any effect on the methyl ester conversion. Acid values of MEs produced from frying oil was 1.16 mgKOH/g when compared with 0.5 mg KOH/g specified as the maximum value according to JUS EN14214 [JUS EN 14214:2004].

4.1.4. Iodine value

The iodine value of the feedstocks used in this study, which is a measure of unsaturation degree, was in the range of 70-78 mg I_2/100 g. According to JUS EN 14214 [JUS EN 14214:2004], MEs used as diesel fuel must have an iodine value less than 120 g I2 per 100 g of sample. Methyl esters obtained in this study had iodine value in the range 72-80g I2/100 g and this finding is in accordance to the fatty acid composition, i.e. the calculated total unsaturation degree of MEs (see Table 4). Iodine value depends on the feedstock origin and greatly influences fuel oxidation tendency. Consequently, in order to avoid oxidation.

4.1.5. Saponification value

The saponification value represents milligrams of potassium hydroxide required to saponify one gram of fat or oil. The obtained results indicated that in general, esters had higher saponification values than the corresponding oils. Saponification values of the feedstocks and products analyzed here, ranged from 199 to 207 mg KOH/g oil. However, knowing that a triglyceride has 3 fatty acid chains associated and each triglyceride will give 3 methyl esters, stoichiometrically it may be expected that the same amount of fatty acid carbon chain in neat feedstock oil and the biodiesel will react with the same amount of KOH giving the soaps, i.e. their saponification values will be the same. But, could this assumption be also applied on the waste frying oils knowing that their properties differ significantly from the neat oils as a consequence of cyclization, polymerization and degradation of fatty acids.

4.1.6. Cetane index

Krisnagkura [1986] proposed the equation for the estimation of cetane index (CI) based on the saponification and iodine values, recommending not to be used for oils, only for methyl esters. Namely, it has been previously documented that despite the fact that triglycerides and fatty acid methyl esters have similar saponification and iodine values, like it was obtained in this study too, cetane indexes of oils are generally much lower than those of methyl ester derivates. Thus, discussion on CI of frying oil will not be made. In this work, the CI value was 38 and this value less than the CI of Egyptian diesel, EN 14214 and D-6751 (55, 51 and 47 respectively). Šiler-Marinkovic´ and Tomaševic [1998] also used CI for the characterization of methyl esters produced from crude frying oils, and the estimated values were from 49.7 to 50.9. As an alternative to cetane number, cetane index is also an indicator of ignition quality of the fuel and is related to the time that passes between injection of the fuel into the cylinder and onset of ignition [Knothe, 2005].

Test	Produced Biodiesel	Egyptian Diesel oil	Biodiesel (EN14214)	Biodiesel D-6751
Flash point °C	202	"/> 55	"/> 101	"/> 130
Density g/cm³ @ 15.56 °C	0.9055	0.82-0.87	0.86-0.9	----
Kinematic Viscosity cSt @ 40 °C	8.38	1.6-7	3.5-5	1.9-6
Kinematic Viscosity cSt @ 100 °C	4.34	---	----	----
Total acid number (mg KOH/g)	0.48	Nil	< 0.5	< 0.8
Cloud point °C	3	---	- 4	----
Pour point °C	0	4.5-15	----	----
Initial boiling point IBP °C	229	----	----	
Cetane number	63.8	Min. 55	"/> 51	"/> 47
Calorific value MJ/Kg	38.54	Min. 44.3	32.9	----
Total S wt%	0.12	Max. 1.2	< 0.01	< 0.05
Ash content wt%	0.002	Max. 0.01	0.02	< 0.02
Carbon residue wt%	0.63	Max.0.1	< 0.03	< 0.05
Copper strip corrosion @ 100°C	1a	1a	Class 1	No. 3 Max.
Water content wt%	0.08	Max. 0.15	< 0.05	0.05
Iodine number mg I₂/100 g	60	---	120	---

Table 3. Physicochemical properties for produced biodiesel compared to the Egyptian standards of petro-diesel fuel and two international biodiesel standards

4.1.7. Fatty acid composition

As can be observed from Table 5, regardless of the fatty acid profiles were observed in the biodiesel produced from frying oil, consisting mainly of methyl esters of oleic (C 18:1), palmitic (C 16:0), and stearic (C 18:0) acids (30.60, 3.0 and 66.40 % respectively) and 2.8 % unknown fatty acid. these results are in agreement with the results obtained by Predojvic (2008) who reported that, fatty acid profiles were observed in the biodiesels produced from sun flower oil consisting mainly of methyl esters of oleic (C 18:1), palmitic (C 16:0), linoleic (C 18:2) and stearic (C 18:0) acids.

Parameters	Feedstock	Produced biodiesel
Acid value mg KOH/g	5.1	0.48
Iodine value mg I_2/g	62.0	60.0
Saponification value mg KOH/g	199.5	207.0

Table 4. Some chemical properties of waste cooking oil (WCO) used as feedstock for methyl esters preparation and produced biodiesel

Fatty acid ester	Carbon number chain	Wt%	Molecular formula
Palmetic	16	3.00	$C_{16}H_{32}O_2$
Stearic	18	66.40	$C_{18}H_{36}O_2$
Oleic	18	30.60	$C_{18}H_{34}O_2$

Table 5. Composition of biodiesel obtained by transesterification of WCO using GC

4.2. Bioethanol

4.2.1. Property of ethanol

Melting point: -114.15

Boiling point: 78.3

Molecular formula: C_2H_5OH

Molecular weight: 46.07

Specific gravity: 0.789

Toxicity: Get intoxicated

4.3. Biomethane

4.3.1. *Gas properties*

4.3.1.1. *Molecular weight*

- Molecular weight : 16.043 g/mol

4.3.1.2. *Solid phase*

- Melting point : -182.5 °C
- Latent heat of fusion (1,013 bar, at triple point) : 58.68 kJ/kg

4.3.1.3. *Liquid phase*

- Liquid density (1.013 bar at boiling point) : 422.62 kg/m³
- Liquid/gas equivalent (1.013 bar and 15 °C (59 °F)) : 630 vol/vol
- Boiling point (1.013 bar) : -161.6 °C
- Latent heat of vaporization (1.013 bar at boiling point) : 510 kJ/kg

4.3.1.4. *Critical point*

- Critical temperature : -82.7 °C
- Critical pressure : 45.96 bar

4.3.1.5. *Gaseous phase*

- Gas density (1.013 bar at boiling point) : 1.819 kg/m³
- Gas density (1.013 bar and 15 °C (59 °F)) : 0.68 kg/m³
- Compressibility Factor (Z) (1.013 bar and 15 °C (59 °F)) : 0.998
- Specific gravity (air = 1) (1.013 bar and 21 °C (70 °F)) : 0.55
- Specific volume (1.013 bar and 21 °C (70 °F)) : 1.48 m³/kg
- Heat capacity at constant pressure (Cp) (1 bar and 25 °C (77 °F)) : 0.035 kJ/(mol.K)
- Heat capacity at constant volume (Cv) (1 bar and 25 °C (77 °F)) : 0.027 kJ/(mol.K)
- Ratio of specific heats (Gamma:Cp/Cv) (1 bar and 25 °C (77 °F)) : 1.305454
- Viscosity (1.013 bar and 0 °C (32 °F)) : 0.0001027 Poise
- Thermal conductivity (1.013 bar and 0 °C (32 °F)) : 32.81 mW/(m.K)

4.3.2. Miscellaneous

- Solubility in water (1.013 bar and 2 °C (35.6 °F)) : 0.054 vol/vol
- Autoignition temperature : 595 °C

5. Biofuel blending

It is important that when you are purchasing fuel you make sure it is high quality by meeting all ASTM specifications. Fuel that is off specification on just one of the ASTM standards can not only cause serious engine problems, but it can void engine warranties if it is determined that the fuel caused damage. This can cause unnecessary costly repairs for vehicles/equipment. To review specifications for diesel fuel, biodiesel and biodiesel blends, see the specifications in the Appendix. In an effort ensure that producers and marketers operate in a manner consistent with proper specifications, the National Biodiesel Accreditation Commission created the BQ-9000 program in 2005. This voluntary program establishes quality systems for producers and marketers of biodiesel in the areas of storage, sampling, testing, blending, shipping, distribution and fuel management practices. If purchasing B100 or a biodiesel blend, ask if the biodiesel is from a BQ-9000 biodiesel producer/marketer. If you are unable to get fuel from a BQ-9000 producer/marketer, the next best thing is to verify with your supplier that the fuel meets all ASTM specifications.

In most cases the blending process takes place right at the terminal rack by a process called in-line blending. This is the preferred method because it ensures complete blending. In-line blending occurs when warm biodiesel is added to a stream of diesel fuel as it travels through a pipe or hose in such a way that the biodiesel and diesel fuel become thoroughly mixed by the turbulent movement. This product is sold directly to customers, petroleum jobbers or a distribution company for sale to customers.

The blend level (percentage of biodiesel in the biodieseldiesel mixture) determines many important characteristics of the blended fuel. A higher-than-specified level of biodiesel may exceed the engine manufacturer's recommended limitation, compromising the engine performance. A lower blend level of biodiesel may reduce the expected benefits, such asfuel lubricity and tail pipe emission. In addition, cloud point and pour point of biodiesel are usually higher than that of diesel fuel, and a higher blend level makes the fuel unsuitable or difficult to use in cold weather conditions. Engine injection timing can be adjusted based on the blend level in order to improve the engine emission and performance (Tat and Van Gerpen, 2003).

It has been reported that the actual biodiesel content of blended biodiesel fuel sold at gas stations can be significantly different from the nominal blend level. A 2% nominal blend has been found to actually contain anywhere from 0% to 8% biodiesel (Ritz and Croudace, 2005). There are several reasons why the actual blend level may differ from the specified level. For instance, if biodiesel is blended at a temperature less than 10°F above its cloud point, it will not mix well with diesel, causing a rich mixture in one portion of the tank and a lean mixture in another portion (NBB, 2005). Other reasons for the discrepancy may include profit-driven fraud and involuntary mixing of diesel into the blend to lower the overall blend level of biodiesel. Biodiesel is usually sold at a higher price than diesel fuel; therefore, the price of the fuel is dependent on the blend level. Knothe (2001) has shown that near-infrared (NIR) spectroscopy and nuclear magnetic resonance (NMR) can be used to detect biodiesel blend levels. However, the NMR method depends on the biodiesel fatty acid profile; hence, knowledge of the biodiesel feedstock is required before this method can be used. In addition, using NMR only to detect blend level may not be cost effective. For NIR spectroscopy, Knothe suggested using wavelengths around 1665 nm or 2083 to 2174_nm. Since aromatic compounds produce strong and sharp infrared bands due to their relatively rigid molecular structure and diesel fuels have varying amounts of aromatics between 20% and 35% (Song et al., 2000), the absorbance of a blend may not directly correlate to the percentage of biodiesel. The absorbance is defined as the logarithm of the radiation intensities ratio, that is, before and after being absorbed by a sample.

Diesel fuel is distilled from crude petroleum, which is composed primarily of hydrocarbons of the paraffinic, naphthenic, and aromatic classes. Each class contains a very broad range of molecular weights. One of the features of diesel fuel is the presence of 20% to 35% aromatic compounds by weight. Aromatics are a class of hydrocarbons that are characterized by a stable chemical ring structure. They are determined primarily by the composition of the crude oil feed, which is usually selected based on considerations of availability and cost (Chevron, 2006). On the other hand, biodiesel is a mixture of fatty acid esters. Fatty acids with 16 to 22 carbon chain lengths are predominant in oils and fats. The resulting mixture of fatty acid esters depends on the kind of feedstock used. Neat biodiesel contains essentially no aromatic compounds.

The presence of aromatics in diesel and their absence in biodiesel creates the possibility of distinguishing these two fuels using ultraviolet spectroscopy. Benzene, the simplest aromatic compound, has maximum absorption at 278 nm (Zawadzki et al., 2007). Biodiesel, which is esters of long-chain fatty acids when adequately diluted in n-heptane, has negligible absorbance compared to the aromatics at the same frequency. Hence, differences in biodiesel feedstocks will have a minimal impact on absorbance at this wavelength. The ultraviolet (UV) range between 200 and 380 nm is also referred to as near-UV. In general, light sources, filters, and detectors are less expensive for this vicinity of the spectrum than for IR at 8621 nm, as used by the CETANE 2000. Hence, near-UV spectroscopy may present a low-cost alternative method for biodiesel blend level sensing (Figure 12 and 13).

6. Material balance of biofuel product

Biomass conversion plant has many components which are connected each other. Material and energy flow among the components, therefore we should grasp the detail of the balance (Figures 13-16). If there is a choke point, the flow stagnation causes to the troubles of operation and low efficiency of the performance. (Masami UENO, University of Ruyku, Faculty of Agriculture, Okinawa, Japan).

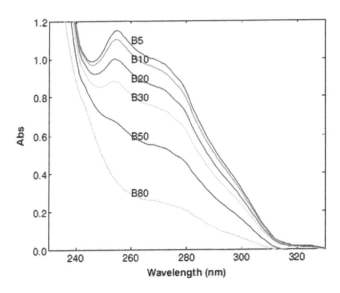

Figure 12. UV absorbance spectra of soy methyl ester and No. 2 diesel blend diluted 1:2915 in *n*-heptane.

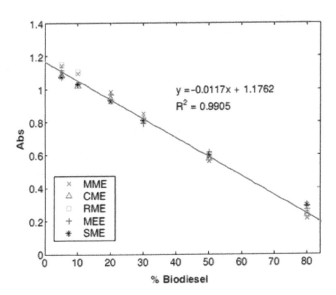

Figure 13. Absorbance of diluted biodiesel-diesel blends from different feedstocks at 260 nm wavelength (MME = mustard methyl esters, CME_=canola methyl esters, RME = rapeseed methyl esters, MEE = mustard ethyl esters, and SME = soybean methyl esters).

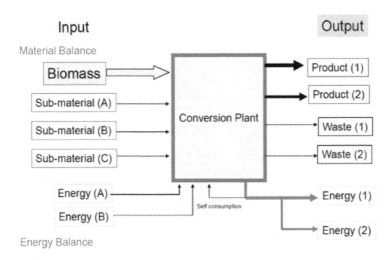

Input

Material Balance

Biomass ⟹	
Sub-material (A) →	
Sub-material (B) →	**Conversion Plant**
Sub-material (C) →	

Energy (A) ──┐
                                    Self consumption
Energy (B) ──┘

Energy Balance

Output

Product (1)
Product (2)
Waste (1)
Waste (2)
Energy (1)
Energy (2)

Figure 14. Material balance and energy balance.

Material and Energy Balances in BDF production

Used cooking oil, Capacity; 5t/d

BDF(5t /d)

Material	kg/d
Used cooking oil	1,000
T-C	806

Material	kg/d
Methanol	108
T-C	41

Material	kg/d
KOH	7

Material	kg/d
A-Heavy oil	11
Heat	493 MJ/d

Type of energy	MJ/d
Electric power	521

BDF plant

Material	kg/d
BDF	1,004
T-C	806.2

Material	kg/d
Glycerin	104
T-C	40.6

Material	kg/d
KOH	7

Type of energy	MJ/d
Heat	493

Figure 15. Material and energy balance in biodiesl fuel production

Material and Energy Balances in Direct Combustion

Wood waste, Capacity; 50t/d

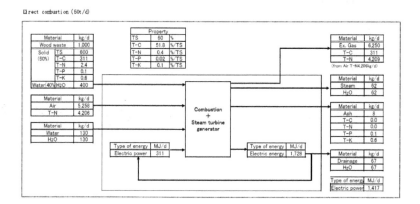

Figure 16. Material and energy balance in direct combustion

Material and Energy Balances in RDF Production

Burnable waste, Capacity; 25t/d

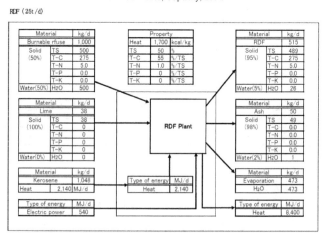

Figure 17. Material and energy balance in RDF production

7. Conclusion

The different kind of biomass considered as main source for biofuel (diesel-methane, ethanol, compost –etc). The cost of extraction and blending is very effective point for use of biomass in addition to ability for use of all part from biomass as multipurpose.

Author details

Emad A. Shalaby

Biochemistry Dept., Facult. of Agriculture, University of Ruyku, Cairo University, Egypt

References

[1] Anonymous. 1992. Chapter 4-Alcohols. In Wine Analysis. Research Institute for Wines, Taiwan Tobacco & Wine Monopoly Bureau. pp. 1-7.

[2] Chevron. 2006. Diesel fuels technical review. San Ramon, Cal.: Chevron Corp.

[3] Collins, T. S., Miller, C. A., Altria, K. D. Waterhouse, A. L. 1997. Development of a rapid method for the analysis of ethanol in wines using capillary electrophoresis. Am. J. Enol. Vitic. 48: 280-284.

[4] Dube'MA.; A.Y. Tremblay, J. Liu, Biodiesel production using a membrane reactor, Bioresour. Technol. 98 (2007) 639–647.

[5] Felizardo P, Neiva Correia MJ, Raposo I, Mendes JF, Berkemeier R, Bordado JM. Production of biodiesel from waste frying oils. Waste Manage 2006;26:487–94.

[6] Hayyan , M., Mjalli , FS., Hashim, MA.; , lNashef, IA. (2010). A novel technique for separating glycerine from palm oil-based biodiesel using ionic liquids. Fuel Processing Technology 91 :116–120.

[7] JUS EN 14214:2004. Automotive fuels. Fatty acid methyl esters (FAME) for diesel engines-requirements and test methods. Standardization Institute,Belgrade, Serbia.

[8] Karaosmanog˘lu F, Cig˘izog˘lu KB, Tuter M, Ertekin S. Investigation of the refining step of biodiesel production. Energ Fuel 1996;10:890–5.

[9] Krisnangkura K. A simple method for estimation of cetane index of vegetable oil methyl esters. JAOCS 1986;3:552–3.

[10] Knothe, G. 2001. Determining the blend level of mixtures of biodiesel with conventional diesel fuel by fiber-optic near-infrared spectroscopy and H-1 nuclear magnetic resonance spectroscopy. J. American Oil Chem. Soc. 78(10): 1025-1028.

[11] Ma F.; M.A. Hanna, Biodiesel production: a review, Bioresour. Technol. 70 (1999) 1–5.

[12] Meher LC, Sagar DV, Naik SN. Techical aspects of biodiesel production by transesterification-a review. Renew Sust Energ Rev 2004:1–21.

[13] Mittelbach M. Diesel fuel derived from vegetable oils, VI: specifications and quality control of biodisel. Bioresour Technol 1996;56:7–11.

[14] Nagai, S. and Nishio, N., In "Handbook of Heat and Mass Transfer, vol. 3 Catalysis, Kinetics and Reactor Engineering" Ed. Cheremisinoff, N.P., 701-752 (1986) Gulf Publishing Corn., Houston, London, Paris, Tokyo.

[15] NBB. 2005. Biodiesel cold weather blending study. Jefferson City, Mo.: National Biodiesel Board.

[16] Predojevic, ZJ. (2008).The production of biodiesel from waste frying oils: A comparison of different purification steps. Fuel 87 : 522–3528.

[17] Ritz, G. P., and M. C. Croudace. 2005. Biodiesel or FAME (fatty acid methyl ester): Mid-infrared determination of ester concentration in diesel fuel. Houston, Tex.: Petroleum Analyzer Company (PAC).

[18] Shalaby, EA (2011). Algal Biomass and Biodiesel Production, Biodiesel - Feedstocks and Processing Technologies, Margarita Stoytcheva and Gisela Montero (Ed.), ISBN: 978-953-307-713-0, InTech, Available from: http://www.intechopen.com/articles/show/title/algal-biomass-and-biodiesel-production

[19] Shalaby, EA and El-Gendy. NS. (2012). Two steps alkaline transesterification of waste cooking oil and quality assessment of produced biodiesel. International Journal of Chemical and Biochemical Sciences. IJCBS, 1(2012): 30-35.

[20] Sharma YC, Singh B, Upadhyay SN. Advancements in development and characterization of biodiesel: a review. Fuel 2008;87:2355–73.

[21] Šiler-Marinkovic' SS, Tomaševic' VA. Transesterification of sunflower oil in situ. Fuel 1998;77:1389–91.

[22] Song, C., C. S. Hsu, and I. Mochida. 2000. Introduction to chemistry of diesel fuel. In Chemistry of Diesel Fuels, 1-60. New York, N.Y.: Taylor and Francis.

[23] Tat, M. E., and J. H. Van Gerpen. 2003. Biodiesel blend detection with a fuel composition sensor. Applied Eng. in Agric. 19(2): 125-131.

[24] Tomasevic V. A. and Siler-Marinkovic S. S. "Methanolysis of used frying oil" Fuel Process Technol. 81: 1-6 (2003).

[25] Vicente G.; M. Martínez, J. Aracil, Optimisation of integrated biodiesel production. Part I. A study of the biodiesel purity and yield, Bioresour. Technol. 98 (2007). 1724–1733.

[26] Zawadzki, A.; Shrestha, D. S.; He, B. (2007). Biodiesel Blend Level Detection Using Ultraviolet Absorption Spectra. American Society of Agricultural and Biological Engineers, 50(4): 1349-1353.

Chapter 3

Gas Fermentation for Commercial Biofuels Production

Fung Min Liew, Michael Köpke and
Séan Dennis Simpson

Additional information is available at the end of the chapter

1. Introduction

With diminishing global reserves of crude oil and increasing demand, especially from developing countries, the pressure on oil supply will grow. Although the 2007-2010 financial crisis brought down the price of crude oil (per barrel) from a record peak of US $145 in July 2008, factors such as recovering global economies and political instability in the Middle East have restored the price of crude oil to the US$100 mark. At current rate of consumption, the global reserves of petroleum are predicted to be exhausted within 50 years [1, 2]. This, coupled with the deleterious environmental impacts that result from accumulating atmospheric CO_2 from the burning of fossil fuels, the development of affordable, and environmentally sustainable fuels is urgently required. Many countries have responded to this challenge by legislating mandates and introducing policies to stimulate research and development (R&D) and commercialization of technologies that allow the production of low cost, low fossil carbon emitting fuels. For instance, the European Union (EU) has mandated member countries to a target of deriving 10% of all transportation fuel from renewable sources by 2020 [3]. Between 2005 and 2010, renewable energies such as solar, wind, and biofuels have been increasing at an average annual rate of 15-50% [4]. Renewable energy accounted for an estimated 16% of global final energy consumption in 2009 [4].

Biofuels have been defined as solid (bio-char), liquid (bioethanol, biobutanol, and biodiesel) and gaseous (biogas, biosyngas, and biohydrogen) fuels that are mainly derived from biomass [5]. Liquid biofuels provided a small but growing contribution towards worldwide fuel usage, accounting for 2.7% of global road transport fuels in 2010 [4]. The world's largest producer of biofuels is the United States (US), followed by Brazil and the EU [4]. In 2009, US and Brazil accounted for approximately 85% of global bioethanol production while Europe generated about 85% of the world's biodiesel [6]. The global market for liquid biofuels (bioethanol and biodiesel) increased dramatically in recent years, reaching US$83 billion in 2011 and is projected to US$139 billion by 2021 [7].

The use and production of biofuels has a long history, starting with the inventors Nikolaus August Otto and Rudolph Diesel, who already envisioned the use of biofuels such as ethanol and natural oils when developing the first Otto cycle combustion and diesel engines [6]. While fermentative production of ethanol has been used for thousands of years, mainly for brewing beer starting in Mesopotamia 5000 B.C., fermentative production of another potential biofuel butanol, has only been discovered over the last century, but had significant impact. During the World War 1, Chaim Weizmann successfully applied a process called ABE (acetone-butanol-ethanol) fermentation using *Clostridium acetobutylicum* to generate industrial scale acetone (for cordites, the propellant of cartridges and shells) from starchy materials [6, 8]. His contribution was later recognised in the Balfour declaration in 1917 and he became the first President of the newly founded State of Israel [6, 8]. Intriguingly, the enormous potential of butanol produced at that time was not realized and the substance was simply stored in huge containers [6]. ABE fermentation became the second biggest ever biotechnological process (after the ethanol fermentation process) ever performed, but the low demand of acetone following the conclusion of the war led to closure of all the plants [8]. Although ABE fermentation briefly made a comeback during the Second World War, increasing substrate costs and increasing stable supply of low cost crude oil from the Middle East rendered the technology economically unviable. Recently, a resurgence of the technology is underway as some old plants are reopened and new plants are being built or planned in China, the US, the United Kingdom (UK), Brazil, France and Austria [6, 8].

Traditionally sugar substrates derived from food crops such as sugar cane, corn (maize) and sugar beet have been the preferred feedstocks for the production of biofuels. However, world raw sugar prices have witnessed significant volatility over the last decade or so, ranging from US$216/ton in year 2000 to a 30 year high of US$795/ton in February 2011 due to global sugar deficits and crop shortfall [9]. This has created uncertainty and raised sustainability issues about its use as a feedstock for large scale biofuel production. This review aims to shed light on the use of syngas and industrial waste gas as feedstocks, and the emerging field of gas fermentation to generate not only biofuels, but also other high-value added products. The advantages of gas fermentation over conventional sugar-based fermentation and thermochemical conversions, and their flexibility in utilizing a spectrum of feedstocks to generate syngas will be discussed. The biochemistry, genetic and energetic background of the microorganisms that perform this bioconversion process will be critically examined, together with recent advances in systems biology and synthetic biology that offer growing opportunities to improve biocatalysts in terms of both the potential products that can be produced and their process performance. The key processes such as gasification, bioreactor designs, media formulation, and product recovery will be analysed. Finally, the state of commercialization of gas fermentation will be highlighted and an outlook will be provided.

2. Advantages of gas fermentation

The production of first generation biofuels relies on food crops such as sugar beet, sugar cane, corn, wheat and cassava as substrates for bioethanol; and vegetable oils and animal fats for bio-diesel. Although years of intense R&D have made methods of bioethanol production (typically using the yeast *Saccharomyces cerevisiae*) technologically mature, there remain some serious questions regarding its sustainability. The use of food crops as a source of carbohydrate feed-stocks by these processes requires high-quality agricultural land. The inevitable conflict be-tween the increasing diversion of crops or land for fuel rather than food production has been highlighted as one of the prime causes of rising global food prices. Furthermore, corn ethanol producers in the US, have historically enjoyed a 45-cent-a-gallon federal tax credit for years (which ended in early 2012), costing the government US$30.5 billion between 2005 to 2011, rais-ing questions about its economic competitiveness with gasoline [10, 11].

These arguments have stimulated the search for so-called second generation biofuels, which utilize non-food lignocellulose biomass such as wood, dedicated energy crops, agricultural residues and municipal solid wastes as feedstocks. Biomass consists of cellulose, hemicellulose and lignin, and the latter of which is extremely resistant to degradation. One approach to un-locking the potential in this abundant feedstock is to separate the lignin from the carbohydrate fraction of the biomass via extensive pre-treatment of the lignocellulose involving, for exam-ple, steam-explosion and/or acid hydrolysis. These pre-treatments are designed to allow the carbohydrate portion of the biomass to be broken down into simple sugars, for example by en-zymatic hydrolysis using exogenously added cellulases to release fermentable sugars [12]. Such approaches have been found to be expensive and rate limiting [6, 12, 13]. Alternatively, processes using cellulolytic microorganisms (such as *C. cellulolyticum*, *C. thermocellum*, and *C. phytofermentans*) to carry out both the hydrolysis of lignocelluloses and sugar fermentation in a single step, termed 'Consolidated Bioprocessing Process (CBP)' [12] have been proposed, how-ever the development of these is still at an early stage, and again low conversion rates seem to be a major limitation that needs to be overcome.

Microorganisms such as acetogens, carboxytrophs and methanogens are able to utilize the CO_2 + H_2, and/or CO available in such syngas as their sole source of carbon and energy for growth as well as the production of biofuels and other valuable products. However, only acetogens are described to synthesize metabolic end products that have potentials as liquid transportation fuels. While biological processes are generally considered slower than chemical reactions, the use of these microbes to carry out syngas fermentation offers several key advantages over alter-native thermo-chemical approaches such as the Fischer-Tropsch' process (FTP). First, microbi-al processes operate at ambient temperatures and low pressures which offer significant energy and cost savings. Second, the ambient conditions and irreversible nature of biological reactions also avoid thermodynamic equilibrium relationships and allow near complete conversion effi-ciencies [14, 15]. Third, biological conversions are commonly more specific due to high enzy-matic specificities, resulting in higher product yield with the formation of fewer by-products. Fourth, unlike traditional chemical catalysts which require a set feed gas composition to yield desired product ratios or suite, microbial processes have freedom to operate for the production of the same suite of products across a wider range of $CO:H_2$ ratios in the feed gas [16]. Fifth, biocatalysts exhibit a much higher tolerance to poisoning by tars, sulphur and chlorine than inorganic catalysts [6, 16]. However, some challenges have been identified for syngas fermentation to be commercialized, including gas mass transfer limitations, long retention times due to slow cell growth, and lower alcohol production rates and broth concentrations. Recent progress and development to remedy these issues will be highlighted in this review.

3. Feedstock and gasification

Due to the flexibility of the microbes to ferment syngas with diverse composition, virtually any carbonaceous materials can be used as feedstock for gasification. Non-food biomass that can be employed as feedstock for gasification includes agricultural wastes, dedicated energy crops, forest residues, and municipal organic wastes, or even glycerol and feathers [16-20]. Biomass is available on a renewable basis, either through natural processes or anthropogenic activities (e.g. organic wastes). It has been estimated that out of a global energy potential from modern biomass of 250 EJ per year in 2005, only 9 EJ (3.6%) was used for energy generation [18]. The use of existing waste streams such as municipal organic waste also differentiate itself from other feedstocks such as dedicated energy crops because these wastes are available today at economically attractive prices, and they are often already aggregated and require less indirect land use. Alternatively, gasification of non-biomass sources such as coal, cokes, oil shale, tar sands, sewage sludge and heavy residues from oil refining, as well as reformed natural gas are commonly applied as feedstocks for the FTP and can also be used for syngas fermentation [15, 21]. Furthermore, some industries such as steel manufacturing, oil refining and chemical production generate large volume of CO and/or CO_2 rich gas streams as wastes. Tapping into these sources using microbial fermentation process essentially convert existing toxic waste gas streams into valuable commodities such as biofuels. The overall process of gas fermentation is outlined in Figure 1.

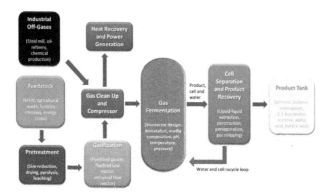

Figure 1. Overview of gas fermentation proces

Prior to gasification, biomass generally needs to go through a pre-treatment process encompassing drying, size reduction (e.g. chipping, grinding and chopping), pyrolysis, fractionation and leaching depending on the gasifier configuration [22, 23]. This upstream pre-treatment process can incur significant capital expense and add to the overall biomass feedstock cost, ranging from US$16-70 per dry ton [22]. Gasification is a thermo-chemical process that converts carbonaceous materials to gaseous intermediates at elevated temperature (600-1000°C), in the presence of an oxidizing agent such as air, steam or oxygen [16, 22]. The resulting syngas contains mainly CO, CO_2, H_2 and N_2, with varying amounts of CH_4, water vapour and trace amount of impurities such as H_2S, COS, NH_3, HCl, HCN, NO_x, phenol, light hydrocarbons and tar [17, 22, 24]. The composition and amount of impurities of syngas depends on the feedstock properties (e.g. moisture, dust and particle size), gasifier type and operational conditions (e.g. temperature, pressure, and oxidant) [17, 22]. Table 1 summarizes typical composition of syngas and other potential gas streams derived from various sources.

4. Fixed bed gasifier

Depending on the direction of the flows of carbonaceous fuel and oxidant (air or steam), fixed bed gasifier can be further categorized into updraft or downdraft reactor. In the updraft (counter-current) version of the fixed bed gasifier, biomass enters from the top while gasifying agent from the bottom. The biomass moves down the reactor through zones of drying (100°C), pyrolysis (300°C), gasification (900°C) and finally oxidation zone (1400°C) [18]. Although this mode of gasifier is often associated with high tar content in the exit gas, recent advances in tar cracking demonstrated that very low tar level is achievable [31]. The direct heat exchange of the oxidizing agent with the entering fuel feed results in low gas exit temperature and hence high thermal efficiency [18, 23]. The downdraft (co-current) gasifier has very similar design as the updraft reactor, except the carbonaceous fuel and oxidizing agent flow in the same direction. In comparison to the updraft gasifier, the downdraft reactor has lower tar content in the exit gas but exhibit lower thermal efficiency [23]. Due to the size limitation in the constriction (where most of the gasification occurs) of the reactor, this mode of gasifier is considered unsuitable for large scale operation [18].

5. Fluidized bed reactor

In fluidized bed reactor, the carbonaceous fuel is mixed together with inert bed material (e.g. silica sand) by forcing fluidization medium (e.g. air and/or steam) through the reactor. The inert bed facilitates better heat exchange between the fuel materials, resulting in nearly isothermal operation conditions and high feedstock conversion efficiencies [18, 22]. The maximum operating temperature of the gasifier is typically around 800 - 900°C, which is limited by the melting point of the bed material [18]. Furthermore, the geometry of the reactor and excellent mixing properties also means that fluidized bed reactors are suitable for up-scaling [18, 22]. Due to these properties, fluidized bed reactor is currently the most commonly used gasifier for biomass feedstock [32]. However, this mode of gasifier is not suitable for feedstocks with high levels of ash and alkali metals because the melting of these components causes stickiness and formation of bigger lumps, which ultimately negatively affect the hydrodynamics of the reactor [18].

	Composition vol%, dry basis						Ref.
	CO	CO$_2$	H$_2$	N$_2$	CH$_4$	Other	
Non-biomass source							
Coal gasification	59.4	10	29.4	0.6	0	0.6	[25]
Coke oven gas	5.6	1.4	55.4	4.3	28.4	4.9	[25]
Partial oxidation of heavy fuel oil	47	4.3	46	1.4	0.3	1	[26]
Hardwood chips + 20 wt%liquid crude glycerol	19.73	11.67	19.38	NR*	3.82	NR*	[19]
Steam reforming of natural gas	15.5	8.1	75.7	0.2	0.5	0	[25]
Steam reforming of Naphtha	6.7	15.8	65.9	2.6	6.3	2.7	[25]
Water gas	30	3.4	31.7	13.1	12.2	9.6	[25]
Steel Mill	44	22	2	32	0	0	[27]
Biomass and organic waste source							
Demolition wood + sewage sludge	10.53	15.02	8.02	60.46	3.19	2.78	[28]
Cacao shell	8	16.02	9.02	61.45	2.34	3.17	[28]
Dairy biomass	8.7	15.7	18.6	56	0.6	0.4	[29]
Switchgrass	14.7	16.5	4.4	56.8	4.2	3.4	[13]
Kentucky bluegrass straw	12.9	17.4	2.6	64.2	2.1	0.8	[30]
Willow	9.4	17.1	7.2	60.42	3.3	2.58	[28]

Note: NR, not reported

The factors that determine which type of gasifier to employ are scale of operation, feedstock size and composition, tar yield and sensitivity towards ash [18]. Currently, three main types of gasifier are commercially employed: fixed bed, fluidized bed and entrained flow reactors [18].

Table 1. Typical composition of syngas and other potential gas streams from various sources

6. Entrained flow reactor

Entrained flow reactor is the preferred route for large scale gasification of coal, petcoke and refinery residues because of high carbon conversion efficiencies and low tar production [22]. This mode of gasifier does not require inert bed material but relies on feeding the feedstocks co-currently with oxidizing agent at high velocity to achieve a pneumatic transport regime

[18]. At operating temperature of 1200-1500°C, this method is able to convert tars and methane, resulting in better syngas quality [18]. Importantly this technology requires the feedstocks to be pulverised into fine particles of ~50 μm before feeding, which is not a major issue for coal but very difficult and costly for biomass sources [18, 22].

7. Microbes and biochemistry of gas fermentation

Acetogens are defined as obligate anaerobes that utilize the reductive acetyl-CoA pathway for the reduction of CO_2 to the acetyl moiety of acetyl-coenzyme A (CoA), for the conservation of energy, and for the assimilation of CO_2 into cell carbon [33]. In addition to the reductive acetyl-CoA pathway, four other biological pathways are known for complete autotrophic CO_2 fixation: the Calvin cycle, the reductive tricarboxylic acid (TCA) cycle, the 3-hydroxypropionate/malyl-CoA cycle and the 3-hydroxypropionate/4-hydroxybutyrate cycle [34]. Since the earlier atmosphere of earth was anoxic and the acetyl-CoA pathway is biochemically the simplest among the autotrophic pathways (the only linear pathway, whereas the other four pathways are cyclic), it has been postulated to be the first autotrophic process on earth [35, 36]. The reductive acetyl-CoA pathway is also known as the 'Wood-Ljungdahl' pathway, in recognition of the two pioneers, Lars G. Ljungdahl and Harland G. Wood, who elucidated the chemical and enzymology of the pathway using *Moorella thermoacetica* (formerly: *Clostridium thermoaceticum*) [35] or CODH/ACS pathway after the key enzyme of the pathway Carbon Monoxide dehydrogenase/Acetyl-CoA synthase. This ancient pathway is diversely distributed among at least 23 different bacterial genera: *Acetitomaculum, Acetoanaerobium, Acetobacterium, Acetohalobium, Acetonema, Alkalibaculum, "Bryantella", "Butyribacterium",Caloramator, Clostridium, Eubacterium, Holophaga, Moorella, Natroniella, Natronincola, Oxobacter, Ruminococcus, Sporomusa, Syntrophococcus, Tindallia, Thermoacetogenium, Thermoanaerobacter,* and *Treponema* [33]. A selection of mesophilic and thermophilic acetogens are presented in Table 2. Acetogens are able to utilize gases CO_2 + H_2, and/or CO to produce acetic acid and ethanol according to the following stoichiometries:

$$2\,CO_2 + 4\,H_2 \Rightarrow CH_3COOH + 2\,H_2O \qquad \Delta H = -75.3\,kJ/mol \tag{1}$$

$$2\,CO_2 + 6\,H_2 \Rightarrow C_2H_5OH + 3\,H_2O \qquad \Delta H = -97.3\,kJ/mol \tag{2}$$

$$4\,CO + 2\,H_2O \Rightarrow CH_3COOH + 2\,CO_2 \qquad \Delta H = -154.9\,kJ/mol \tag{3}$$

$$6CO + 3\,H_2O \Rightarrow C_2H_5OH + 4\,CO_2 \qquad \Delta H = -217.9\,kJ/mol \tag{4}$$

The Acetyl-CoA pathway is essentially a terminal electron-accepting process that assimilates CO_2 into biomass [35]. It constitutes an *Eastern* (or Carbonyl) branch and a *Western* (or Meth-

Species	Substrate	Product(s)	T_{opt} (°C)	pH_{opt}	Genome Status	Ref.
Mesophilic Microorganisms						
Acetobacterium woodii	H_2/CO_2, CO	Acetate	30	6.8	Available	[41, 42]
Acetonema longum	H_2/CO_2	Acetate, butyrate	30-33	7.8		[43]
Alkalibaculum bacchi	H_2/CO_2, CO	Acetate, ethanol	37	8.0-8.5		[44, 45]
Blautia producta	H_2/CO_2, CO	Acetate	37	7		[46]
Butyribacterium methylotrophicum	H_2/CO_2, CO	Acetate, ethanol, butyrate, butanol	37	6		[47-49]
Clostridium aceticum	H_2/CO_2, CO	Acetate	30	8.3	Under construction	[50-52]
Clostridium autoethanogenum	H_2/CO_2, CO	Acetate, ethanol, 2,3-butanediol, lactate	37	5.8-6.0		[27, 53]
Clostridium carboxidivorans or "P7"	H_2/CO_2, CO	Acetate, ethanol, butyrate, butanol, lactate	38	6.2	Draft	[54, 55]
Clostridium drakei	H_2/CO_2, CO	Acetate, ethanol, butyrate	25-30	5.8-6.9		[55-57]
Clostridium formicoaceticum	CO	Acetate, formate	37	NR		[50, 58, 59]
Clostridium glycolicum	H_2/CO_2	Acetate	37-40	7.0-7.5		[60, 61]
Clostridium ljungdahlii	H_2/CO_2, CO	Acetate, ethanol, 2,3-butanediol, lactate	37	6	Available	[27, 62, 63]
Clostridium magnum	H_2/CO_2	Acetate	30-32	7.0		[64, 65]
Clostridium mayombei	H_2/CO_2	Acetate	33	7.3		[66]
Clostridium methoxybenzovorans	H_2/CO_2	Acetate, formate	37	7.4		[67]
"Clostridium ragsdalei" or "P11"	H_2/CO_2, CO	Acetate, ethanol, 2,3-butanediol, lactate	37	6.3		[68]
Clostridium scatologenes	H_2/CO_2, CO	Acetate, ethanol, butyrate	37-40	5.4-7.5		[55, 56]
Eubacterium limosum	H_2/CO_2, CO	Acetate	38-39	7.0-7.2	Available	[41, 69]
Oxobacter pfennigii	H_2/CO_2, CO	Acetate, butyrate	36-38	7.3		[70]
Thermophilic Microorganisms						
Moorella thermoacetica	H_2/CO_2, CO	Acetate	55	6.5-6.8	Available	[71-73]
Moorella thermoautotrophica	H_2/CO_2, CO	Acetate	58	6.1		[74]
Thermoanaerobacter kiuvi	H_2/CO_2	Acetate	66	6.4		[72]

Notes: NR, not reported

Table 2. Acetogens

yl) branch (Figure 2.). The *Western* branch employs a series of enzymes to carry out a six-electron reduction of CO_2 to the methyl group of acetyl-CoA, starting from the conversion of CO_2 to formate by formate dehydrogenase. Formyl-H_4folate synthase then condenses formate with H_4folate to form 10-formyl-H_4folate, which is then converted to 5,10-methenyl-H_4folate by a cyclohydrolase. This is followed by a dehydrogenase that reducesmethenyl- to 5,10-methylene-H_4hydrofolate, before (6S)-5-CH_3-H_4folate is formed by methylene-H_4folate reductase [37]. A B12-depedent methyltransferase (MeTr) then transfer the methyl group of (6S)-5-CH_3-H_4folate to corrinoid iron-sulphur protein (CoFeSP) of the bi-functional carbon monoxide dehydrogenase/acetyl-CoA synthase (CODH/ACS) complex [37]. The bi-functional CODH/ACS enzyme complex is formed by two autonomous proteins, an $\alpha_2\beta_2$ tetramer (CODH/ACS) and a $\gamma\delta$ heterodimer (CoFeSP), and the genes are often arranged in an operon, together with MeTr [37, 38]. In the *Eastern* branch, the CODH component catalyzes the reduction of CO_2 to CO. The central molecule, acetyl-CoA, is finally generated when CO, methyl group (bound to CoFeSP) and CoASH are condensed by ACS. Given the pivotal role of CODH/ACS, it is unsurprising that this complex was found to be the most highly expressed transcripts under autotrophic conditions in *C. autoethanogenum* [27], and can represent up to 2% of the soluble cell protein of an acetogen [39]. CODH/ACS is not unique to acetogenic bacteria, as it is also present in sulphate-reducing bacteria, desulfitobacteria, and Archaea (methanogens and Archaeoglobus) [38, 40].

The reducing equivalents required for fixation of CO_2 carbon into acetyl-CoA come from the oxidation of molecular hydrogen under chemolithoautotrophic growth, or NADH and reduced ferredoxin under heterotrophic growth [75]. An extensive review by Calusinska *et al.* (2010) highlighted the diversity of ubiquitous hydrogenases that Clostridia possess although only one acetogen *C. carboxidivorans* was included in this study [76], which catalyze the reversible oxidation of hydrogen:

$$H_2 \Leftrightarrow 2\,H^+ + 2\,e^- \qquad\qquad (5)$$

The direction of the hydrogenase reaction is directed by the redox potential of the components able to interact with the enzyme. Hydrogen evolution occurs when electron donor is available, whereas the presence of electron acceptor results in hydrogen oxidation [77]. Hydrogenases can be classified into three phylogenetically distinct classes of metalloenzymes: [NiFe]-, [FeFe]-, and [Fe]-hydrogenases [76]. In *Methanosarcina barkeri*, the Ech hydrogenase, a [NiFe]-hydrogenase, was demonstrated to oxidize H_2 to reduce ferredoxin [78]. During acetoclastic methanogenesis, Ech hydrogenase oxidize ferredoxin to generate H_2 [78]. Although genome analysis revealed the presence of Ech-like hydrogenase in *C. thermocellum*, *C. phytofermentans*, *C. papyrosolvens*, and *C. cellulolyticum*, their physiological roles remained unknown [76]. Clostridia harbour multiple distinct [FeFe]-hydrogenases, perhaps reflecting their ability to respond swiftly to changing environmental conditions [76]. The monomeric, soluble [FeFe]-hydrogenase of *C. pasteurianum* is one of the best studied. It transfer electrons from reduced ferredoxins or flavodoxins to protons, forming H_2 [79]. A trimeric [FeFe]-hydrogenase found in *C. difficile*, *C. beijerinckii*, and *C. carboxidivorans* were hypothesized to couple formate oxidation to reduce protons into H_2 [76]. In *Thermotoga maritima*, an electron bifurcating, trimeric [FeFe]-hydrogenase was identified, that was shown to simultaneously oxidize reduced ferredoxin and NADH to evolve hydrogen under low H_2 partial pressure [80]. Under high H_2 partial pressure, the authors hypothesized that the NADH is oxidized to produce ethanol. *In silico* analysis revealed homologs of this bifurcating hydrogenase in a few Clostridia including *C. beijerinckii* and *C. thermocellum* [80]. In addition to classical hydrogenases, CODH/ACS and pyruvate:ferredoxin oxidoreductase (PFOR) from *M. thermoacetica* were shown to have hydrogen evolving capability, possibly as a mean of disposing excess reducing equivalents when electron carriers are limited and/or CO concentration is sufficient to inhibit conventional hydrogenases [81].

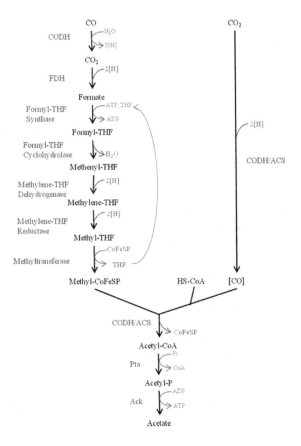

Figure 2. Wood-Ljungdahl Pathway. Ack, acetate kinase; ACS, acetyl-CoA synthase; CODH, carbon monoxide dehydrogenase; CoFeSP, corrinoid iron sulfur protein; FDH, formate dehydrogenase; Pta, phosphotransacetylase; THF, tetrahydrofolate.

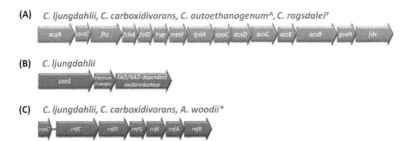

Figure 3. The organization of genes involved in acetogenesis and energy conservation from sequenced key aceto-gens. (A) Wood-Ljungdahl cluster; (B) carbon monoxide dehydrogenase (CODH) cluster; (C) Rnf complex cluster. *acsA*, CODH subunit; *acsB*, ACS subunit; *acsC*, corrinoid iron-sulfur protein large subunit; *acsD*, corrinoid iron-sulfur protein small subunit; *acsE*, methyltransferase subunit; *cooC*, gene for CODH accessory protein; *cooS*, CODH; *fchA*, formimido-tetrahydrofolate cyclodeaminase; *fdx*, ferredoxin; *fhs*, formyl-tetrahydrofolate synthase; *folD*, bifunctional methylene-tetrahydrofolate dehydrogenase/formyl-tetrahydrofolate cyclohydrolase; *gcvH*, gene for glycine cleavage system H protein; *hyp*, hypothetical protein; *lpdA*, dihydrolipoamide dehydrogenase; *metF*, methylene-tetrahydrofolate reduc-tase; *rnfA, rnfB, rnfC, rnfD, rnfE, rnfG*, electron transport complex protein subunits; *rseC*, sigma E positive regulator. ^, truncated *acsA*. #, truncated *fdx*. *, lack *rseC*.

Most acetogens are also able to utilize another gas carbon monoxide (CO). In contrast to CO_2, CO can serve as both a source of carbon btut also as source of electrons such that hy-drogen is not necessarily required. With a CO_2/CO reduction potential of -524 to -558mV, CO is approximately 1000-fold more capable of generating extremely low potential electrons than NADH, capable of reducing cellular electron carriers such as ferredoxin and flavodoxin [38, 82]. The reducing equivalents generated from CO oxidation can be coupled to reduction of CO_2 into acetate, butyrate and/or methane, evolution of molecular hydrogen from pro-tons, reduction of nitrate/nitrite, reduction of sulfur species and reduction of aldehydes into alcohols [35, 83]. However, relatively few microorganisms are able to utilize CO as sole car-bon and energy source, probably due to growth inhibition from sensitivity of their metallo-proteins and hydrogenases towards CO [38, 83]. During exponential growth of *Pseudomonas carboxydovorans* (an aerobic carboxydotroph), it was demonstrated via immunological locali-zation studies that 87% of the key enzyme CODH is associated with the inner cytoplasmic membrane, but this association was lost at the end of the exponential growth phase and a reduction in CO-dependent respiration rate was observed [84, 85]. It should be mentioned that aerobic and anaerobic CODH enzymes are structurally very different. CODH has been reported to be a very rapid and efficient CO oxidizer at rates between 4,000 and 40,000 s^{-1}, and reduces CO_2 at $11s^{-1}$ [86, 87]. Other electron donors commonly used by acetogens in-clude formate, CH_3Cl, lactate, pyruvate, alcohols, betaine, carbohydrate, acetoin, oxalate and citrate [88]. CODH is able to split water in a biological water-gas shift reaction into hydro-gen and electron according to the stoichiometry:

$$CO + H_2O \Rightarrow CO_2 + 2H^+ + 2e^- \qquad (6)$$

The operation of this water gas shift reaction is the biochemical basis for the tremendous flexibility that acetogens have in terms of input gas composition. Via this reaction these organisms can flexibly use CO or H_2 as a source of electrons.Recently, some acetogens such as *C. ljungdahlii, C. aceticum, M. thermoacetica, Sporomusa ovata,* and *S. sphaeroides* have additionally been shown to utilize electrons derived from electrodes to reduce CO_2 into organic compounds such as acetate, formate, fumarate, caffeine, and 2-oxo-butyrate [89]. Termed microbial electrosynthesis, this nascent concept offers another route for acetogens to harvest the electrons generated from sustainable sources (e.g. solar and wind) to reduce CO_2 into useful multi-carbon products such as biofuels [90].

Under chemolithoautotrophic conditions, acetogenesis must not only fix carbon but also conserve energy. Approximately 0.1 mol of ATP is required for generation of 1g of dry biomass in anaerobes [82]. Acetyl-CoA is an energy rich molecule that through the combined actions of Pta (phosphotransacetylase) and Ack (acetate kinase), one ATP can be generated via substrate level phosphorylation (SLP). However, the activation of formate to 10-formyl-H4folate in the methyl-branch of Acetyl-CoA pathway consumes one ATP so no net gain in ATP is achieved via this mechanism [35, 75]. Furthermore, the reduction of CO_2 to the carbonyl group also requires energy, estimated at one third of ATP equivalent [35]. Recent advances indicated that other modes of energy conservation such as electron transport phosphorylation (ETP) or chemiosmotic processes that are coupled to the translocation of protons or sodium ions are implicated in acetogens. Acetogens such as *M. thermoacetica* harbour membrane-associated electron transport system containing cytochrome, menaquinones, and oxidoreductases that translocate H^+ out of the cell [33]. For acetogens that lack such membranous electron transport system, such as *Acetobacterium woodii* and *C. ljungdahlii*, a membrane-bound corrinoid protein is hypothesized to facilitate extrusion of Na^+ or protons during the transfer of methyl group from methyl-H_4F to CODH/ACS [75]. However, all enzymes involved are predicted to be soluble rather than membrane bound. Recent evidence suggested coupling to an Rnf complex in *A. woodii*, and *C. ljungdahlii* (Figure 3) which acts as ferredoxin:NAD⁺-oxidoreductase [62, 91-93]. The Rnf complex is also found in other Clostridia (but not in ABE model organism *C. acetobutylicum*) and bacteria, and was originally discovered in *Rhodobacter capsulatus* where it is involved in nitrogen fixation [93]. Using reduced ferredoxin (Fd^{2-}) generated from CO oxidation, carbohydrate utilization and/or hydrogenase reactions, this membrane-bound electron transfer complex is predicted to reduce NAD^+ with concomitant translocation of Na^+/H^+. The ion gradient generated from the above processes is harvested by H^+- or Na^+- ATP synthase to generate ATP [33, 93]. The recent genome sequencing of *A. woodii* revealed that Rnf complex is likely to be the only ion-pumping enzyme active during autotrophic growth and the organism's entire catabolic metabolism is optimized to maximize the Fd^{2-}/NAD^+ ratio [42]. Recently, a third mechanism of energy conservation which involves bifurcation of electrons by hydrogenases was proposed for anaerobes [94] and demonstrated for enzymes hydrogenase (see above; [80]), butyryl-CoA dehydrogenase [94, 95], or an iron-sulfur flavoprotein Nfn [96]. A similar mechanism has also been proposed for the methylene-THF reductase of the reductive acetyl-CoA pathway, which would enable this highly exergonic reduction step ($\Delta G^{0'} = -22$ kJ/mol) to be coupled with the Rnf complex for additional energy conservation [62]. However, no experimental proof to support this hypothesis has been published to date.

In an attempt to generate an autotrophic *E. coli*, the genes encoding MeTr, the two subunits of CODH/ACS, and the two subunits of CoFeSP from *M. thermoacetica* were cloned and heterologously expressed in *E. coli* [97]. Although the MeTr was found to be active, the other subunits misassembled hence no active enzymes were found [97]. Autotrophic capability is clearly a very complex process that involves many genes other than the CODH/ACS complex and tetrahydrofolate pathway, including compatible cofactors, electron carriers, specific chaperones and energy conservation mechanisms. For instance, more than 200 genes are predicted to be involved in methanogenesis and energy conservation from CO_2 and H_2 in methanogens [98]. A recent patent application described the introduction of three Wood-Ljungdahl pathway genes encoding MeTr, CoFeSP subunit α and β from *C. difficile* into *C. acetobutylicum* [99]. The recombinant strain was shown to incorporate more CO_2 into extracellular products than wild-type [99].

8. Products of gas fermentation

Acetyl-CoA generated via the Wood-Ljungdahl pathway serves as key intermediate for synthesis of cell mass as well as products. All acetogens are described to produce acetate, in order to gain energy via SLP to compensate for the energy invested in activating formate in the *Western* branch of the reductive acetyl-CoA pathway. Acetate and ATP are formed via acetyl-phosphate through the successive actions of Pta and Ack. *pta* and *ack* are arranged in the same operon and they were reported to be constitutively expressed [100]. With CO_2 and H_2 as substrate, only acetate has been observed as major product [44], with minor amounts of ethanol produced in rare cases with *C. ljungdahlii* [101], *C. autoethanogenum* [53], or *"Moorella sp."* [102, 103]. Using the more reduced substrate CO, production of a range of other products have been reported, such as ethanol, butanol, butyrate, 2,3-butanediol [104], and lactate (Figure 4.) [105]. From a biofuel perspective, ethanol and butanol are of particular interest. Ethanol and butanol have even been described as the main fermentation products over acetate in some acetogens under specific conditions. Ethanol producers include *C. ljungdahlii* [62, 63], *C. autoethanogenum* [53], *"C. ragsdalei"* (*"Clostridium* strain P11") [106, 107], *"Moorella sp."* [102, 103], *Alkalibaculum bacchii* [44], *C. carboxidivorans* (*"Clostridium* strain P7") [54, 55], and *B. methylotrophicum* [49, 108]. The latter two have also been described to produce butanol.

Due to historical roles in ABE fermentation, organisms like *C. acetobutylicum*, *C. beijerinckii*, *C. saccharobutylicum*, and *C. saccharoperbutylacetonicum* have been much more extensively characterized than acetogenic Clostridia [95]. Since *C. acetobutylicum* was the first *Clostridium* to be fully sequenced [109] and it remains the most commonly used species for industrial production of solvents to date [110], it provides a model for study of solventogenesis. Although sugar- and starch-utilizing ABE Clostridia and acetogens exhibit clear distinctions in substrate utilization and thus metabolism, they share some similarities in the biochemical pathway and genetic organization of product synthesis and can be used as model for comparison. Structure of key genes and operons (except for the absence of acetone biosynthetic genes) have been found to be very similar in sequenced acetogen *C. carboxidivorans* [54], and in respect of acetate and ethanol genes to some extent also in *C. ljungdahlii* [62]. For instance, the operon structure of *pta-ack, ptb-buk* and the *bcs* cluster of acetogen *C. carboxidivorans* are highly similar to starch-utilizing *C. acetobutylicum* and *C. beijerinckii* [54, 109] (Figure 5). Due to these reasons, solventogenic genes from starch-utilizing Clostridia are ideal targets for heterologous expression in acetogens for improvement of product yield and expansion of product range.

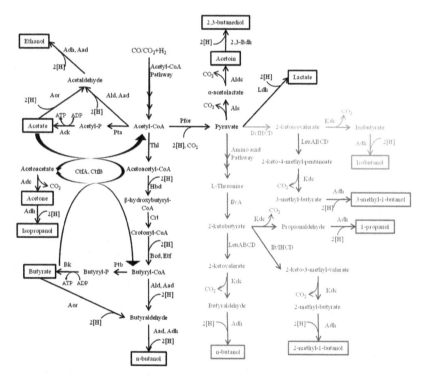

Figure 4. Scheme of metabolite production from gas fermentation using native and genetically modified Clostridia. Black denotes well-characterized pathways in Clostridia. Blue shows demonstrated heterologous pathways that have been engineered into Clostridia. Purple designates hypothetical pathways that can be engineered into Clostridia. Products are highlighted in boxes. Aad, aldehyde/alcohol dehydrogenase; Ack, acetate kinase; Adc, acetoacetate decarboxylase; Adh, alcohol dehydrogenase; Ald, aldehyde dehydrogenase; Aldc, acetolactate decarboxylase; Aor, aldehyde oxidoreductase; Bcd, butyryl-CoA dehydrogenase; Bk, butyrate kinase; Crt, crotonase; CtfA & CtfB, CoA transferase A & B; Etf, electron-transferring flavoprotein; Hbd, hydroxybutyryl-CoA dehydrogenase; IlvA, threonine deaminase; IlvIHCD, valine and isoleucine biosynthesis; Kdc, 2-ketoacid decarboxylase; Ldh, lactate dehydrogenase; LeuABCD, leucine and norvaline biosynthesis; Pfor, Pyruvate ferredoxin oxidoreductase; Pta, phosphotransacetylase; Ptb, phosphotransbutyrylase; Thl, thiolase; 2,3-Bdh, 2,3-butanediol dehydrogenase.

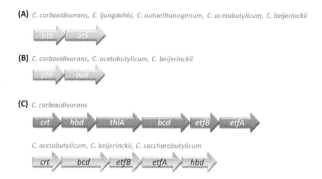

Figure 5. Similarity of acidogenesis and butanol formation gene clusters of acetogens and sugar-utilizing Clostridia. (A) Acetate-forming operon; (B) butyrate-forming operon; (C) butanol-forming operon. *ack*, acetate kinase; *buk*, butyrate kinase; *bcd*, butyryl-CoA dehydrogenase; *crt*, crotonase; *etfA*, electron-transferring flavoprotein subunit A; *etfB*, electron-transferring flavoprotein subunit B; *hbd*, 3-hydroxybutyryl-CoA dehydrogenase; *ptb*, phosphotransbutyrylase; *thlA*, thiolase.

Similar to sugar- and starch-utilizing ABE Clostridia, acetogens such as *C. carboxidivorans* [111, 112], *C. ljungdahlii* [113], and *C. autoethanogenum* [27] also typically undergo biphasic fermentation under autotrophic conditions. The first phase involves the production of carboxylic acids (acidogenic), H_2 and CO_2 during exponential growth. This is followed by the solventogenic phase in which part of the produced acids are reassimilated or reduced into solvents, which usually occurs during stationary growth phase [114]. This shift from acidogenesis to solventogenesis is of industrial importance and several transcriptional analysis on *C. acetobutylicum* [100, 115], and *C. beijerinckii* [116] have been performed to shed light on this process. In both organisms, the onset of solventogenesis coincides with an increase in expression of master sporulation/solventogenesis regulator gene *spo0A*, solventogenic genes such as *ald*, *ctfA-ctfB*, and *adc*, as well as down-regulation of chemotaxis/ motility genes [100, 115, 116]. Physiologically, the signals that induce solventogenesis were hypothesized to involve temperature, low pH, high concentrations of undissociated acetic and butyric acids, limiting concentrations of sulphate or phosphate, ATP/ADP ratio and/or NAD(P)H levels [117].

For Clostridia such as acetogen *C. carboxidivorans* [54], which harbour the genes thiolase (*thlA*), 3-hydroxybutyryl-CoA dehydrogenase (*hbd*), crotonase (*crt*) and butyryl-CoA dehydrogenase (*bcd*), the two carbon acetyl-CoA can be converted to four carbon butyryl-CoA [95]. ThlA compete with the activities of Pta, Ald (aldehyde dehydrogenase), and PFOR to condense two acetyl-CoA into one acetoacetyl-CoA, and plays a key role in regulating the C2:C4 acid ratio [110, 118]. Since the formation of acetate yields twice as much ATP per mole of acetyl-CoA relative to butyrate formation, thiolase activity indirectly affects ATP yield [118]. Under physiological conditions, Crt catalyzes dehydration of β-hydroxybutyryl-CoA to crotonyl-CoA [119]. Bcd was shown to require a pair of electron transfer flavoproteins (EtfA and EtfB) to convert crotonyl-CoA to butyryl-CoA [120]. Furthermore, the Bcd was demonstrated to form a stable complex with EtfA and EtfB, and they were shown to couple the reduction of crotonyl-CoA to butyryl-CoA with concomitant generation of reduced ferre-doxins, which can be used for energy conservation via Rnf complex [94, 119]. Subsequent actions of phosphotransbutyrylase (*ptb*) and butyrate kinase (*buk*) then generate ATP and butyrate from butyryl-CoA [118].

Under low extracellular pH of 4-4.5, the secreted undissociated acetic acid (pK_a 4.79) and/or butyric acid (pK_a 4.82) diffuse back into cell cytoplasm and then dissociate into the respec-tive salts and protons because of the more alkaline intracellular conditions. Without further interventions, the result of this is abolishment of the proton gradient and inevitable cell death [95]. The conversion of acetate and butyrate into solvents increase the pH, thus pro-vide some time for the organism to sporulate and secure long term survival. However, the solvents produced are toxic because they increase membrane fluidity and disrupt critical membrane-associated functions such as ATP synthesis, glucose uptake and other transport processes [114, 121]. In *C. acetobutylicum*, it has been demonstrated that the addition of 7-13 g/l of butanol, or up to 40 g/l of acetone and ethanol resulted in 50% growth inhibition [122]. The bacterium is likely to experience a different cytotoxic effect from endogenously pro-duced solvents because the organism has time to adapt to increasing amount of solvents.

The reassimilation of acetate and butyrate into the respective acyl-CoA and acetoacetate is catalyzed by acetoacetyl-CoA:acetate/butyrate CoA transferase (CtfA and CtfB) [110, 117, 118]. Acetoacetate is deconstructed by acetoacetate decarboxylase (Adc) into acetone and CO_2. This enzyme is missing in acetogenic *C. carboxidivorans* compared to the ABE strains [54, 123]. Some ABE strains such as *C. beijerinckii* NRRL B593 also possess a primary/secon-dary alcohol dehydrogenase that converts acetone to isopropanol [124]. In acetogenic "*C. ragsdalei*", reduction of acetone to isopropanol was also observed although the mechanism of this reduction is as yet unknown [124, 125]. Again, *C. carboxidivorans* lacks this activity [125]. The recycled acetyl-CoA and butyryl-CoA can be converted to ethanol and butanol through the actions of coenzyme A-acylating aldehyde dehydrogenase (Ald) and alcohol de-hydrogenase (Adh) [110, 118]. Ald converts acyl-CoA into aldehydes, and the enzyme has been purified from *C. beijerinckii* NRRL B593 and was shown to be NADH-specific, exhibit higher affinity with butyraldehyde than acetaldehyde, but possess no Adh activity [126]. In *C. ljungdahlii*, two variants of aldehyde:ferredoxin oxidoreductases (AOR) are present in the genome, and they are hypothesized to couple reduced ferredoxin from CO oxidation via the CODH (see above) to perform the reversible reduction of acetate into acetaldehyde, which can be further reduced into ethanol [62].

The final step of solventogenesis utilizes Adh to reduce acetaldehyde and butyraldehyde in-to ethanol and butanol, respectively. For ethanol synthesis, transposon mutagenesis and en-zymatic assay in *C. acetobutylicum* showed the involvement of a specific Ald that does not interact with butyryl-CoA, and a NAD(P)H-dependent Adh [127, 128]. The production of butanol by *C. acetobutylicum* is mainly due to the action of butanol dehydrogenase A and B (BdhA and BdhB), and bifunctional butyraldehyde/butanol dehydrogenase 1 and 2 (AdhE1 and AdhE2) [95]. In *C. carboxidivorans* [54] and *C. ljungdahlii* [62] both *adhE1* and *adhE2* are arranged in tandem and separated by a 200bp gap which contains a putative terminator [62,

111]. This is likely the result of gene duplication [62]. qRT-PCR analysis from *C. carboxidivorans* fed with syngas showed that the two *adhE* showed differential expression, and the more abundant *adhE2* was significantly upregulated over 1000 fold in a time span that coincided with the greatest rate of butanol production [111].

Pyruvate is a central molecule for anabolism and it is predominantly generated from glycolysis during heterotrophic growth. But under autotrophic growth, this four carbon molecule can be synthesized by PFOR and potentially also the pyruvate-formate lyase (PFL). Two variants of PFOR were reported in *C. autoethanogenum*, and transcriptional analysis showed that they were differentially expressed when grown using industrial waste gases (containing CO, CO_2 and H_2) [104]. Unlike PFL from most other microorganisms that only catalyze the lysis of pyruvate into formate and acetyl-CoA, clostridial PFL (*C. kluyveri*, *C. butylicum*, and *C. butyricum*) were reported to readily catalyze the reverse reaction (i.e. pyruvate formation) [129]. Apart from roles in anabolism, pyruvate is also a precursor to other products such as lactic acid and 2,3-butanediol. Small amounts of lactic acid are converted from pyruvate in acetogens, a reaction which is catalyzed by lactate dehydrogenase (Ldh) [104, 118]. Recently, Köpke *et al.* (2011) reported the production of 2mM 2,3-butanediol from acetogenic bacteria (*C. autoethanogenum*, *C. ljungdahlii*, and *C. ragsdalei*) using industrial waste gases (containing CO, CO_2 and H_2) as feedstock [104]. Pyruvate is first converted into α-acetolactate by the enzyme acetolactate synthase, followed by acetolactate decarboxylase which split acetolactate into acetoin and CO_2, before a final reduction of acetoin into 2,3-butanediol by 2,3-butanediol dehydrogenase [104] (Figure 4).

9. Strain improvement and metabolic engineering

The genomes of several solventogenic Clostridia, including gas fermenting species, have been sequenced since 2001 [54, 62, 109, 119, 123, 130], and an array of transcriptomic [100, 116, 121, 131, 132], proteomic [132] and systems analysis [133, 134] are being made increasingly available. However, the generation of stable recombinant Clostridia has been severely hindered by the difficulties encountered introducing foreign DNA into cells and a lack of established genetic tools for this genera of bacteria. In comparison to starch-utilizing Clostridia, very little information is available for metabolic engineering of acetogens. Although this section describes recent advances in the development of genetic tools for mostly sugar-utilizing Clostridia, these techniques are highly relevant and applicable to the closely related acetogenic Clostridia for biofuels or chemical production via gas fermentation.

The ideal microbial catalyst for industrial scale gas fermentation might exhibit the following traits: high product yield and selectivity, low product inhibition, no strain degeneration, asporogenous, prolonged cell viability, strong aero-tolerance, high biomass density and efficient utilization of gas substrates. These can be achieved by directed evolution, random mutagenesis and/or targeted genetic engineering. Traditionally, chemical mutagenesis [135-137] and adaption strategies [138, 139] have been deployed to select for these traits. However, these strategies are limited and often come with the expense of unwanted events.

First attempts of targeted genetic modification of Clostridia were made in the early 1990s by the laboratory of Prof. Terry Papoutsakis [140-142]. While these pioneering efforts relied on use of plasmids for (over)expression of genes in *C. acetobutylicum*, more sophisticated tools were later developed for a range of solventogenic and pathogenic Clostridia.

Antisense RNA (asRNA) has been employed to down-regulate genes. Here, single stranded RNA binds to a complementary target mRNA and prevents translation by hindering ribosome-binding site interactions [143]. For instance, this method has been used to knockdown *ctfB* resulting in production of 30 g/l solvents with significantly suppressed acetone yield in *C. acetobutylicum* ATCC 824 [144, 145].

Several homologous recombination methods have been developed for integration or knock-out of genes in a range of sugar-utilizing Clostridia. In early stage, knockout mutants were almost exclusively generated from single crossover events that could revert back to wild-type [146-152], with stable double crossovers only observed in rare cases [153, 154]. For *C. acetobutylicum* [155] and cellulolytic *C. thermocellum* [156] counter selectable markers have been developed to allow more efficient screening for the rare second recombination event.

ClosTron utilizes the specificity of mobile group II intron Ll.*ltrB* from *Lactoccocus lactis* to propagate into a specified site in the genome via a RNA-mediated, retro-homing mechanism which can be used to disrupt genes [157]. This technique has initially been developed by InGex and Sigma-Aldrich under the name 'TargeTron™' and successfully adapted to a range of solventogenic and pathogenic Clostridia including *C. acetobutylicum*, *C. difficile*, *C. sporogenes*, *C. perfringens*, and *C. botulinum* [158-160] by the laboratory of Prof. Nigel Minton.

The same laboratory recently also developed another method for integration of DNA into the genome. Termed Allele-Coupled Exchange (ACE), this approach does not employ a counter selective marker to select for the rare second recombination event. Rather, it utilizes the activation or inactivation of gene(s) that result in a selectable phenotype, and asymmetrical homology arms to direct the order of recombination events [161]. Remarkably, the whole genome of phage lambda (48.5kb minus a 6kb region) was successfully inserted into the genome of *C. acetobutylicum* ATCC 824 in three successive steps using this genetic tool. This technique was also demonstrated in *C. difficile* and *C. sporogenes* [161].

For reverse engineering, mainly transposon mutagenesis has been utilized. Earlier efforts of transposon mutagenesis were demonstrated in *C. acetobutylicum* P262 (now: *C. saccharobutylicum* [162]), *C. acetobutylicum* DSM792, *C. acetobutylicum* DSM1732, and *C. beijerinckii* NCIM 8052, but issues with multiple transposon insertions per mutant, and non-random distribution of insertion were reported [163, 164]. Recent developments have seen the successful generation of mono-copy random insertion of transposon *Tn1545* into cellulolytic *C. cellulolyticum* [165] and mariner transposon *Himar1* into pathogenic *C. difficile* [166].

While there is still a lack of some other essential metabolic engineering tools such as efficient inducible promoters, the array of available tools that enabled significant improvements to the ABE process and cellulolytic Clostridia fermentations as summarized in Table 3.

Organism	Genetic modification	Phenotypes/Effects	Ref
Acetogens			
C. ljungdahlii	Plasmid overexpression of butanol biosynthetic genes from C. acetobutylicum (thlA, crt, hbd, bcd, adhE and bdhA)	Produced 2 mM butanol from syngas	[62]
C. autoethanogenum	Plasmid overexpression of butanol biosynthetic genes from C. acetobutylicum (thlA, crt, hbd, bcd, etfA, & etfB)	Produced 26 mM butanol using steel mill gas	[167]
C. autoethanogenum	Plasmid expression of native groES and groEL	Increased alcohol tolerance	[168]
C. aceticum	Plasmid overexpression of acetone operon from C. acetobutylicum (adc, ctfAB, thlA)	Produced up to 140 µM acetone using gas	[169, 170]
Acidogenesis and Solventogenesis			
C. acetobutylicum	Inactivation of buk and overexpression of aad	Produced same amount of butanol as control but relatively more ethanol, corresponding to a total alcohol tolerance of 21.2 g/l	[171]
C. acetobutylicum	Inactivation of hbd using ClosTron	Produced 716 mM ethanol by diverting C4 products	[172]
C. acetobutylicum	Inactivation of ack using ClosTron	Reduction in acetate kinase activity by more than 97% resulted in 80% less acetate produced but similar final solvent amount	[173]
C. tyrobutylicum	Inactivation of ack and plasmid overexpression of adhE2 from C. acetobutylicum	Produced 216 mM butanol	[174]
C. thermocellum	Inactivation of ldh and pta via homologous recombination	Showed 4 fold increase in ethanol yield (122 mM instead of 28 mM)	[156]
C. cellulolyticum	Inactivation of ldh and mdh (malate dehydrogenase) using ClosTron	Generated 8.5 times higher ethanol yield (56.4 mM) than wild type (6.5 mM)	[175]
C. acetobutylicum	Plasmid overexpression of a synthetic acetoneoperon (adc, ctfA, ctfB) and primary/secondary adh from C. beijerinckii NRRL B593	Produced 85 mM isopropanol	[176]

C. acetobutylicum	Genome insertion of *adh* gene from *C.beijerinckii* NRRL B593 using Allele-Coupled Exchange	Converted acetone into 28 mM isopropanol without affecting the yield of other fermentation products	[161]
Biosynthesis of New Products			
C. cellulolyticum	Plasmid overexpression of *kivD*, *yqhD*, *alsS*, *ilvC* and *ilvD*	Produced 8.9 mM isobutanol by diverting 2-ketoacid intermediates	[177]
C. acetobutylicum	Plasmid expression of native *ribGBAH* operon and mutated PRPP amidotransferase	Produced 70 mg/l riboflavin and 190 mM butanol	[178]
Solvent- and Aero-tolerance			
C. acetobutylicum	Plasmid overexpression of glutathione *gshA* and *gshB* from *E. coli*	Improved aero- and solvent-tolerance	[179]
C. acetobutylicum	Plasmid overexpression of chaperone *groESL*	Showed 85% decrease in butanol inhibition and 33% increase in solvent yield	[180]
Substrate Utilization			
C. acetobutylicum	Plasmid expression of *acsC, acsD* and *acsE* from *C. difficile*	Increased incorporation of CO_2 into extracellular products	[99]
C. saccharoperbutylacetonicum strain N1-4	Knockdown hydrogenase *hupCBA* expression using siRNA delivered from plasmid	Significantly reduced hydrogen uptake activity to 13% (relative to control strain)	[181]

Table 3. Genetically modified solventogenic Clostridia

In contrast, to date only a limited number of acetogenic Clostridia have been successfully modified. Pioneering work in this area has been undertaken in the laboratory of Prof. Peter Dürre. C. *ljungdahlii*, a species that does not naturally produce butanol, was modified with butanol biosynthetic genes (*thlA, hbd, crt, bcd, adhE* and *bdhA*) from C. *acetobutylicum* ATCC 824 resulting in production of up to 2 mM of butanol using synthesis gas as sole energy and carbon source [62]. By delivering a plasmid with acetone biosynthesis genes *ctfA, ctfB, adc,* and *thlA* in C. *aceticum,* production of up to 140 μM acetone was demonstrated from various gas mixes (80% H_2/20% CO_2 and 67% H_2/33% CO_2) [169, 170]. Recent patent filings by Lanza-Tech describe the production of butanol as main fermentation product and increased alcohol tolerance in genetically engineered acetogens. Up to 26 mM butanol were produced with genetically modified C. *ljungdahlii* and C. *autoethanogenum* using steel mill gas (composition 44% CO, 32% N_2, 22% CO_2, and 2% H_2) as the only source of carbon and energy when the butanol biosynthetic genes *thlA, hbd, crt, bcd, etfA,* and *etfB* were heterologously expressed [167]. Overexpression of native *groESL* operon in C. *autoethanogenum* resulted in a strain that displayed higher alcohol tolerance relative to wild-type when challenged with ethanol [168].

Besides the classical Clostridial butanol pathway (which constitutes genes *thlA, crt, hbd, bcd, etfA and etfB*; see earlier section), a non-fermentative approach has been described and demonstrated in *E. coli* for branched chain higher alcohol production [182]. This alternative approach requires a combination of highly active amino acid biosynthetic pathway and artificial diversion of 2-keto acid intermediates into alcohols by introduction of two additional genes: broad substrate range 2-keto-acid decarboxylase (*kdc*) which converts 2-keto acids into aldehydes, followed by *Adh* to form alcohols [182]. Engineered strains of *E. coli* have been shown to produce alcohols such as isobutanol, n-butanol, 2-methyl-1-butanol, 3-methyl-1-butanol and 2-phenylethanol via this strategy [182]. For instance, the overexpression of *kivD* (KDC from *Lactococcus lactis*), *adh2, ilvA,* and *leuABCD* operon, coupled with deletion of *ilvD* gene and supplementation of L-threonine, increased n-butanol yield to 9 mM while producing 10 mM of 1-propanol [182].An even more remarkable yield of 300 mM isobutanol was achieved through introduction of *kivD, adh2, alsS* (from *B. subtilis*), and *ilvCD* into *E. coli* [182]. Like butanol, isobutanol exhibits superior properties as a transportation fuel when compared to ethanol [177]. By applying similar strategy into *C. cellulolyticum*, 8.9 mM isobutanol was produced from cellulose when *kivD, yqhD, alsS, ilvC,* and *ilvD* were overexpressed [177]. This result suggests that such non-fermentative pathway is suitable target for metabolic engineering of acetogens for the biosynthesis of branched chain higher alcohols. Via synthetic biology and metabolic engineering, production of additional potential liquid transportation fuels like farnesese or fatty acid based fuels has successfully been demonstrated in *E. coli* or yeast from sugar [183, 184]. Given the unsolved energetics in acetogens, it is unclear if production of such energy dense liquid fuels could be viable via gas fermentation.

10. Fermentation and product recovery

10.1. Bioreactor design

An optimum gas fermentation system requires efficient mass transfer of gaseous substrates to the culture medium (liquid phase) and microbial catalysts (solid phase). Gas-to-liquid mass transfer has been identified as the rate-limiting step and bottleneck for gas fermentation because of the low aqueous solubility of CO and H_2, respectively at only 77% and 68% of that of oxygen (on molar basis) at 35°C [185]. Hence, a bioreactor design that delivers sufficient gas-to-liquid mass transfer in an energy-efficient manner at commercial scale for gas fermentation represents a significant engineering challenge. A brief overview of reactor configurations reported in gas fermentation operations is given below.

10.2. Continuous Stirred Tank Reactor (CSTR)

In continuous stirred tank reactor (CSTR), gas substrates are continuously fed into the reactor and mechanically sheared by baffled impellers into smaller bubbles, which has greater interfacial surface area for mass transfer [16]. In addition, finer bubbles have a slower rising velocity and a longer retention time in the aqueous medium, resulting in higher gas-to-liquid mass transfer [24]. Fermentation reactions using *C. ljungdahlii* have been successfully maintained in a 2 litre CSTR under autotrophic conditions for more than a month, while achieving peak ethanol level of 6.5 g/l and CO conversion rate of 93% [186]. The production of 49 g/l of ethanol from gas substrates using *C. ljungdahlii* was demonstrated using CSTR [113]. In another example, a 100 litre stirred tank reactor was demonstrated to produce up to 24.57 g/l ethanol, 9.25 g/l isopropanol and 0.47 g/l n-butanol during a 59-day semi-batch gas fermentation using "*C. ragsdalei*" strain P11 as biocatalysts [112]. An improved version of CSTR incorporates microbubble sparger to generate finer bubbles to achieve higher mass transfer coefficient [187]. Although CSTR offers complete mixing and uniform distribution of gas substrates to the microbes, the high power per unit volume required to drive the stirrer are thought to make this approach economically unviable for commercial scale gas fermentation systems [187].

10.3. Bubble column reactor

In contrast to CSTR, gas mixing in bubble column reactor is achievable by gas sparging, without mechanical agitation. This reactor configuration has fewer moving parts, and consequently has a lower associated capital and operational costs while exhibiting good heat and mass transfer efficiencies, making it a good candidate for large scale gas fermentation [17]. However, excessive level of gas inflow for enhanced mixing have been cited as an issue that leads to heterogeneous flow and back-mixing of the gas substrates [16, 17]. *C. carboxidivorans* strain P7 was cultured in a 4 litre bubble column reactor for 20 days using a combination of producer gas and synthetic syngas, generating a peak ethanol concentration of 6 g/l [13].

10.4. Immobilized cell column reactor

One of the key challenges of gas fermentation is cell density. Immobilization of microbes through crosslinking or adsorption to insoluble biosupport materials and the subsequent packing within the column offers a range of benefits [14]. These include high cell densities, plug flow operation, high mass transfer rate via direct contact between microbe and gas, reduction of retention time, and operation without mechanical agitation [14, 16]. However, channelling issues may arise when the microbe overgrows and completely fill the interstitial space. Due to limitations in column dimensions and packing, this reactor configuration lacks flexibility to operate in various gas fermentation conditions [14, 16].

10.5. Trickle-bed reactor

Trickle-bed reactor is a gas- or liquid- continuous reactor consisting of packed bed, which liquid culture trickles down through packing media containing suspended or immobilized cells [16, 24, 187]. The gas substrate is delivered either co-currently or counter-currently to the liquid flow, and no mechanical agitation is required [187].In this reactor format, low gas and liquid flow rates are typically applied, generating relatively low pressure drops [187]. Trickle-bed reactor was found to exhibit excellent gas conversion rates and higher productivities than CSTR and bubble column reactor [15].

11. Gas fermentation parameters

11.1. Gas composition

The gas composition and its impurities can have an impact on the productivity of the gas fermentation process. Greater molar ratio of H_2:CO allows greater efficiency in the conversion of the carbon from CO into products such as ethanol, because reducing equivalents are generated from oxidation of H_2 (rather than CO). However, CO is also a known inhibitor of hydrogenase which can affect utilization of H_2 during fermentation. In *B. methylotrophicum*, H_2 utilization was inhibited until CO was exhausted [108]. When CO is consumed, acetogens are able to grow using CO_2 and H_2. Common impurities from biomass gasification or other waste gases are tar, ash, char, ethane, ethylene, acetylene, H_2S, NH_3 and NO [17, 22, 24, 188].These have been shown to cause cell dormancy, inhibition of hydrogen uptake, low cell growth and shift between acidogenesis and solventogenesis in acetogens [13, 188]. For instance, NH_3 from the feed gas readily convert into NH_4^+ in the culture media and these ions were recently shown to inhibit hydrogenase and cell growth of acetogen "*C. ragsdalei*" [189]. A number of strategies to mitigate the impact of such impurities have been proposed, for example installing 0.025 mm filters, or the use of gas scrubbers or cyclones, and improvement in gasification efficiency and scavenging for contaminants in the gas stream using agents such as potassium permanganate, sodium hydroxide or sodium hypochlorite [24, 190-192]. H_2S does not have a negative effect on acetogens such as *C. ljungdahlii* up to 5.2% (v/v) [193].

11.2. Substrate pressure

The partial pressure of syngas components have a major influence on microbial growth and product profiles because the enzymes involved are sensitive to substrate exposure [194]. Due to the low solubility of CO and H_2 in water, the growth of dense bacterial cell cultures can face mass transfer limitations, so increasing the partial pressure of gaseous substrates can help alleviate this problem. For instance, studies in which the CO partial pressure (P_{CO}) increased from 0.35 to 2.0 atm showed that this resulted in a 440% increase in maximum cell density, a significant increase in ethanol productivity and a decrease in acetate production in *C. carboxidivorans* strain P7 [195]. In another study involving *C. ljungdahlii*, the increase of P_{CO} from 0.8 to 1.8 atm had a positive effect on ethanol production, and the microbe did not exhibit any substrate inhibition at high P_{CO} [196].In less CO-tolerant microorganisms, the effect of increasing P_{CO} partial pressure range from non-appreciable in the case of *Rhodospirillum rubrum* [197], to negative impact on doubling time of *Peptostreptococcus productus* (now: *Blautia product*) [194] and *Eubacterium limosum* [198]. Similar to CO, the increase in partial pressure of H_2 (pH_2) to 1700 mbar enhanced acetate productivity of *A. woodii* to 7.4g acetate/ l/day [199].

11.3. Medium formulation

Although acetogens are able of utilizing CO and CO_2/H_2 as carbon and energy source, other constituents such as vitamins, trace metal elements, minerals and reducing agents are also required for maintenance of high metabolic activity [16, 113]. Studies indicated that formation of ethanol in solventogenic Clostridiais non-growth associated and limitation of growth by reducing availability of carbon-, nitrogen- and phosphate- nutrients shift the balance from acidogenesis to solventogenesis [113, 200, 201]. Optimization of medium formulation for *C. ljungdahlii* through reduction of B-vitamin concentrations and elimination of yeast extract significantly enhanced the final ethanol yield to 48 g/l in a CSTR with cell recycling (23 g/l without cell recycling) [113]. Another study by Klasson *et al.* showed thatthe replacement of yeast extract with cellobiose not only increased maximum cell concentration, but also enhanced ethanol yield by 4-fold [14]. Media formulation for *C. autoethanogenum* was investigated using Plackett-Burman and central composite designs, but only low ethanol yield was recorded overall [202]. In an attempt to reduce the cost of fermentation medium and improve process economics, 0.5 g/l of cotton seed extract without other nutrient supplementation was shown to be a superior medium for *C. carboxidivorans* strain P7 in producing ethanol from syngas fermentation [203]. A recent study showed that increasing concentrations of trace metal ions such as Ni^{2+}, Zn^{2+}, SeO_4^-, WO_4^-, Fe^{2+} and elimination of Cu^{2+} from medium improved enzymatic activities (FDH, CODH, and hydrogenase), growth and ethanol production in "*C. ragsdalei*" under autotrophic conditions [107].

A low redox potential is necessary for strict anaerobes to grow, hence reducing agents such as sodium thioglycolate, ascorbic acid, methyl viologen, benzyl viologen, titanium (III)–citrate, potassium ferricyanide, cysteine-HCl and sodium sulfide are commonly added to fermentation medium [14, 16, 204]. Furthermore, the addition of reducing agent directs the electron and carbon flow towards solventogenesis by enhancing the availability of reducing equivalents to form NADH for alcohol production [16, 205]. Excessive addition of reducing agents can cause slower microbial growth due to reduced ATP formation from acetogenesis so it is important to determine the optimum concentration of reducing agents [14, 16]. The sulfur containing gases (e.g. H_2S) present in syngas are toxic to chemical catalysts but can be beneficial for microbial catalysts by reducing medium redox potential, stimulate redox sensitive enzymes such as CODH, and promote alcohol formation [206, 207].

11.4. Medium pH

Like other organisms, acetogens have a limited range of pH for optimal growth so the pH of the fermentation medium needs to be closely controlled. The extracellular pH directly influences the intracellular pH, membrane potential, proton motive force, and consequently substrate utilization and product profile [208, 209]. In most studies, lowering pH medium divert carbon and electron flow from cell and acid formation towards alcohol production [113, 209-211]. By applying this knowledge, Gaddy and Clausen performed a two-stage CSTR syngas fermentation systems using *C. ljungdahlii* where they set the first reactor at pH 5 to promote cell growth, and pH 4 - 4.5 in the second reactor to induce ethanol production [212]. One recent study with *C. ljungdahlii* showed conflicting results in which cell density and ethanol production were both higher at pH 6.8 when compared to pH 5.5 [213].

11.5. Temperature

The optimum temperature for mesophilic acetogens are between 30-40°C, while thermophilic acetogens grow best between 55 and 58°C. The fermentation temperature not only affects substrate utilization, growth rate and membrane lipid composition of the acetogens, but also gas substrate availability because gas solubility increases with decreasing temperature [24, 211]. "*C. ragsdalei*"was reported to produce more ethanol at 32°C than at the optimum growth temperature of 37°C [211].

12. Cell separation and product recovery

To retain high cell densities in reactor, microbes can be grown as biofilm attached to carrier material. Planktonic cells can be retained in the fermentation broth by installing solid/liquid separators such as membranous ultra-filtration units, spiral wound filtration systems, hollow fibres, cell-recycling membranes and centrifuges [214-216]. The concentrations of solvents from gas fermentation rarely exceed 6% [w/v] so a cost- and energy- efficient product recovery process is required. Furthermore, acetogens also exhibit low resistance towards solvents like ethanol [217, 218] and butanol [219, 220] so an *in situ*/online product recovery system can enhance solvent productivity by decreasing solvent concentrations (and hence toxicity) in the fermentation broth. Distillation has been the traditional method of product recovery but the associated high energy costs have led to the development of alternative methods such as liquid-liquid extraction, pervaporation, perstraction, and gas stripping [24, 221].

12.1. Liquid-liquid extraction

In liquid-liquid extraction, a water-insoluble organic extractant is mixed with the fermentation broth [222]. Because solvents are more soluble in the organic phase than in the aqueous phase, they get selectively concentrated in the extractant.Although this technique does not remove water or nutrients from the fermentation broth, some gaseous substrates might be removed because CO and H_2 have much higher solubility in organic solvents than water [222, 223]. Oleyl alcohol has been the extractant of choice due to its relatively non-toxicity [224].

12.2. Perstraction

Liquid-liquid extraction is associated with several problems including toxicity to the microbes, formation of emulsion, and the accumulation of microbes at the extractant and fermentation broth interphase [222]. In an attempt to remediate these problems, perstraction was developed and this technique employs membrane to separate the extractant from the fermentation broth. This physical barrier prevent direct contact between the microbe and the toxicity of extractant, but it can also limit the rate of solvent extraction and is susceptible to fouling [219, 221]

12.3. Pervaporation

In a product recovery technique termed pervaporation, a membrane that directly comes in contact with fermentation broth is used to selectively remove volatile compounds such as ethanol and butanol [219, 222]. The volatile compounds diffuse through the membrane as vapour and are then collected by condensation. To facilitate volatilization of permeates into vapour, a partial pressure difference across the membrane is usually maintained by applying a vacuum or inert gas (e.g. N_2) across the permeate side of the membrane [219]. Polydimethylsiloxane (PDMS) is the current material of choice for the membrane, but other materials such as poly(1-trimethylsilyl-1-propyne) (PTMSP), hydrophobic zeolite membranes, and composite membranes have also been investigated [225].

12.4. Gas stripping

Gas stripping is an attractive product recovery method for gas fermentation because the exit gas stream from the bioreactor can be used for *in situ*/online product recovery [219]. Following product recovery via condensation, the effluent and gas can be recycled back into the bioreactor. In sugar-based fermentation using *C. beijerinckii* mutant strain BA101, *in situ* gas stripping was shown to improve ABE productivity by 200%, complete substrate utilization and also complete acid conversion into solvents, when compared to non-integrated process [226].

13. Commercialization

The growing commercial interests in using gas fermentation as a platform for biofuels production is evident in the recent spike in patent fillings within the field [105]. A 2009 report compared mass and energy conversion efficiencies from a process engineering standpoint between enzymatic hydrolysis fermentation of lignocellulose, syngas fermentation and FTP [227]. The authors concluded that while syngas fermentation offers a range of advantages such as low pretreatment requirement and low energy requirement for bioconversion, the technology is severely limited by low ethanol productivity [227]. Another report documented the techno-economic analysis of gas fermentation and concluded that the selling price of ethanol using this technology would still be significantly higher than gasoline in 2009 [228]. In contrast, Griffin and Schultz recently compared the production of ethanol from CO-rich gas using thermo-chemical route and biological gas fermentation route [22]. The authors concluded that gas fermentation offers superior fuel yield per volume of biomass feed, carbon conversion to fuel, energy efficiency and lower carbon emissions relative to the thermo-chemical approach to bioethanol production.

Ethanol and butanol are the most attractive fuel products from current gas fermentation but other by-products such as 2,3-butanediol, acetic acid and butyric acid are also valuable commodities that have the potential to provide significant additional revenue streams, setting off costs for biofuel production. 2,3-butanediol is a high value commodity which can be used to synthesize chemical products such as 1,3-butanediane, methyl ethyl ketone, and gamma butyrolactone, with a combined potential market value of $43 billion [104]. Acetic acid is an important precursor for synthesis of polymers while butyric acid can be used as a flavouring agent in the food industry [229, 230]. With the development of advanced genetic tools for expansion of product range, the industry might witness an increasing emphasis on the production of high-value commodities in addition to biofuels.

Several companies are actively engaged in the development of the gas fermentation technology and some are approaching commercialization. Bioengineering Resources Inc (BRI) founded by Prof. James Gaddy of University of Arkansas, Fayetteville, an early pioneer in the investigation of gas fermentation at scale, was the first company to explore the potential of gas fermentation for industrial bioethanol production. BRI was acquired by chemical company INEOS and rebranded as INEOS Bio (www.ineosbio.com). A pilot-scale facility in Arkansas has been operated since 2003 using several isolates of *C. ljungdahlii* [231] and is building a US$130 million commercial facility in Florida with its joint venture partner New Planet Energy Florida [232]. The commercial facility is expected to start operation in the second quarter of 2012 and is aiming to generate 8 million gallon of cellulosic ethanol per annum and 6 MW of power to the local communities [232]. INEOS Bio also announced design of a second plant, the Seal Sands Biorefinery in Teeside, UK [233].

Founded in 2006, Coskata Inc. (www.coskata.com) is a US-based company that has reported achieving ethanol yields of 100 gallons per dry ton of wood biomass in a semi-commercial facility in Pennsylvania [234]. The company licensed several microbial strains from the University of Oklahoma [235], which has filed patents and journal publications for acetogens such as "*C. ragsdalei*" [211, 236, 237] and *C. carboxidivorans* [55, 112]. A patent documenting a new ethanologenic species, "*C. coskatii*" was also recently filed by Coskata [238]. Backed by a conditional US$250 million loan guarantee from the US Department of Agriculture (USDA), Coskata has announced that it is planning to build a commercial plant with the capacity to produce 55 million gallon fuel grade ethanol per annum in Alabama [234, 239]. While the initial strategy saw biomass as feedstock, the company recently announced its first commercial plant will be switched to 100% natural gas as feedstock [240]. A planned IPO with the aim to tap into private investors to finance the plant was put on hold [241]. In 2012, Coskata and INEOS Bio were involved in a trade secret dispute which culminated in a settlement that see INEOS Bio receiving US$2.5 million cash payment, shares and right to receive 2.5% of future ethanol royalties from Coskata [242].

LanzaTech is a NZ/US based company that has developed a gas fermentation technology to utilize industrial off-gases from steel making and other sources, as well as syngas produced from biomass as feedstocks. The company has reported the development of a proprietary Clostridial biocatalyst that is able to convert the CO-rich waste gas with minimal gas conditioning into bioethanol and the platform chemical 2,3-butanediol. The use of industrial off-gases as feedstock not only helps to reduce the carbon footprint of the steelmaking operations but also allows the production of valuable commodities without the costs associated with feedstock gasification. The company has estimated that up to 30 billion gallon of bioethanol per year can be produced from the CO-rich off gases produced through steel manufacturers globally [243]. Founded in 2005, LanzaTech has successfully demonstrated bioethanol production at a pilot plant at BlueScope Steel in Glenbrook, NZ, since 2008 and the company has recently started operating its 100,000 gallon bioethanol per year demonstration facility in Shanghai, China, using waste gas collected from an adjacent steel mill plant owned by its partner Baosteel Group [243, 244]. LanzaTech is planning to build a commercial facility with the capacity to produce 50 million gallon of bioethanol per annum in China by 2013 [243]. The recent acquisition of a biorefinery facility developed by the US-based gasification technology company Range Fuels in Georgia, and a milestone signing of its first commercial customer, Concord Enviro Systems (India), highlighted LanzaTech's intention to utilize MSW and lignocellulosic waste as feedstocks for biofuel and chemical production [243, 244].

14. Conclusion

One of the fundamental factors that govern the environmental and economical sustainability of biofuel production is feedstock. Through gasification, a spectrum of renewable non-food feedstock such as agricultural wastes, dedicated energy crops, forest residues, and MSW can be converted into syngas. This article presents a detailed examination of gas fermentation technology in capturing the carbon and energy from syngas and produce biofuels and chemicals. In comparison to indirect fermentation of lignocellulose via enzymatic hydrolysis, and thermo-chemical FTP, gas fermentation offers several advantages such as good product yield and selectivity, operation in ambient conditions, high tolerance to gas impurities, and elimination of expensive pre-treatment steps and costly enzymes. Furthermore, some industries such as steel mill, natural gas steam reforming, oil refining and chemical production generate large volumes of CO-rich off-gas. Gas fermentation can access these existing feedstocks and generate valuable products from these while reducing carbon emissions. Pivotal to gas fermentation is acetogens such as *C. ljungdahlii*, *C. carboxidivorans*, "*C. ragsdalei*" and *C. autoethanogenum*, which are able to metabolize CO, and CO_2/H_2 into a range of products such as ethanol, butanol, isopropanol, acetone, 2,3-butanediol, acetic acid and butyric acid. Sustained effort in studying the physiology and biochemistry using advanced molecular techniques such as genomics, transcriptomics, proteomics, metabolomics and systems biology are essential to further the understanding of these microbes. Furthermore, recent advances in Clostridial genetic tools offer endless opportunities to engineer strains that have improved product yield, substrate utilization, no strain degeneration, and synthesis of new products.

The main challenges associated with commercialization of gas fermentation have been identified as gas-to-liquid mass transfer limitation, product yield, substrate utilization efficiency, low biomass density and product recovery. Further development of bioreactor is necessary to improve the availability of gas substrates and maintain high cell density for higher productivity. Improvement in integrated product recovery technology is also essential to lower the costs of product recovery and alleviate product inhibition. Gas fermentation appears to be mature enough for commercialization since several companies have already demonstrated their technologies at pilot scale and are moving towards commercialization in the near future.

Author details

Fung Min Liew, Michael Köpke and Séan Dennis Simpson

LanzaTech NZ Ltd., Parnell, Auckland, New Zealand

References

[1] Vasudevan PT, Gagnon MD, Briggs MS. Environmentally sustainable biofuels – The case for biodiesel, biobutanol and cellulosic ethanol. In: Singh OV, Harvey SP, editors. Sustainable Biotechnology: Sources of Renewable Energy: Springer Science +Business Media B.V.; 2010. p. 43-62.

[2] Appenzeller T. The end of cheap oil. National Geographic Magazine. 2004.

[3] European_Union. Directive 2009/28/EC of the European Parliament and of the Council of 23 April 2009 on the promotion of the use of energy from renewable sources and amending and subsequently repealing Directives 2001/77/EC and 2003/30/EC. 2009.

[4] REN21. Renewables 2011 Global Status Report. Paris: REN21, 2011.

[5] Demirbas A. Political, economic and environmental impacts of biofuels: A review. Applied Energy. 2009 Nov;86:S108-S17. PubMed PMID: WOS:000271170300013.

[6] Köpke M, Noack S, Dürre P. The past, present, and future of biofuels – Biobutanol as Promising Alternative. In: dos Santos Bernades MA, editor.: InTech; 2011. p. 451-86.

[7] Pernick R, Wilder C, Winnie T. Clean Energy Trends 2012. Clean Edge Inc., 2012.

[8] Dürre P. Biobutanol: an attractive biofuel. Biotechnology journal. 2007;2(12):1525-34.

[9] OECD-FAO. OECD-FAO Agricultural Outlook 2011-2020. 2011.

[10] Murse T. Understanding the ethanol subsidy 2011. Available from: http://usgovinfo.about.com/od/moneymatters/a/The-Federal-Ethanol-Subsidy.htm.

[11] Pear R. After three decades, tax credit for ethanol expires. The New York Times. 2nd January 2012.

[12] Carere CR, Sparling R, Cicek N, Levin DB. Third generation biofuels via direct cellulose fermentation. International Journal of Molecular Sciences. 2008 Jul;9(7):1342-60. PubMed PMID: WOS:000257986800014.

[13] Datar RP, Shenkman RM, Cateni BG, Huhnke RL, Lewis RS. Fermentation of biomass-generated producer gas to ethanol. Biotechnology and bioengineering. 2004;86(5):587-94.

[14] Klasson KT, Ackerson MD, Clausen EC, Gaddy JL. Bioreactor design for synthesis gas fermentations. Fuel. 1991 May;70(5):605-14. PubMed PMID: WOS:A1991FL49400009.

[15] Klasson KT, Ackerson MD, Clausen EC, Gaddy JL. Bioconversion of synthesis gas into liquid or gaseous fuels. enzyme and microbial technology. 1992 Aug;14(8):602-8. PubMed PMID: WOS:A1992JE44900001.

[16] Mohammadi M, Najafpour GD, Younesi F, Lahijani P, Uzir MH, Mohamed AR. Bioconversion of synthesis gas to second generation biofuels: A review. Renewable & Sustainable Energy Reviews. 2011 Dec;15(9):4255-73. PubMed PMID: WOS: 000298764400005.

[17] Abubackar HN, Veiga MC, Kennes C, Coruña L. Biological conversion of carbon monoxide: rich syngas or waste gases to bioethanol. Biofuels, Bioproducts & Biorefining. 2011:93-114.

[18] Siedlecki M, de Jong W, Verkooijen AHM. Fluidized bed gasification as a mature and reliable technology for the production of bio-syngas and applied in the production of liquid transportation fuels-A review. Energies. 2011 Mar;4(3):389-434. PubMed PMID: WOS:000288788000002.

[19] Wei L, Pordesimo LO, Haryanto A, Wooten J. Co-gasification of hardwood chips and crude glycerol in a pilot scale downdraft gasifier. Bioresource Technology. 2011 May; 102(10):6266-72. PubMed PMID: WOS:000291125800097.

[20] Dudynski M, Kwiatkowski K, Bajer K. From feathers to syngas - Technologies and devices. Waste Management. 2012 Apr;32(4):685-91. PubMed PMID: WOS: 000302822500007.

[21] Sipma J, Henstra AM, Parshina SM, Lens PN, Lettinga G, Stams AJM. Microbial CO conversions with applications in synthesis gas purification and bio-desulfurization. Critical reviews in biotechnology. 2006;26(1):41-65.

[22] Griffin DW, Schultz MA. Fuel and chemical products from biomass syngas: A comparison of gas fermentation to thermochemical conversion routes. Environmental Progress & Sustainable Energy. 2012 Jul;31(2):219-24. PubMed PMID: WOS: 000302794000008.

[23] McKendry P. Energy production from biomass (part 3): gasification technologies. Bioresource Technology. 2002 May;83(1):55-63. PubMed PMID: WOS:000175355300006.

[24] Munasinghe PC, Khanal SK. Biomass-derived syngas fermentation into biofuels: Opportunities and challenges. Bioresour Technol. 2010;101(13):5013-22.

[25] Bartish CM, Drissel GM. Carbon monoxide. In: Othmer K, editor. Encyclopedia of Chemical Technology. New York: John Wiley and Sons; 1978.

[26] van Houten RT, Lettinga G. Biological sulphate reduction with synthesis gas: Microbiology and technology. Immobilized Cells: Basics and Applications. 1996;11:793-9. PubMed PMID: WOS:A1996BF78D00106.

[27] Köpke M, Mihalcea C, Liew FM, Tizard JH, Ali MS, Conolly JJ, et al. 2,3-butanediol production by acetogenic bacteria, an alternative route to chemical synthesis, using industrial waste gas. Applied and Environmental Microbiology. 2011 Aug;77(15): 5467-75. PubMed PMID: WOS:000293224500045.

[28] van der Drift A, van Doorn J, Vermeulen JW. Ten residual biomass fuels for circulating fluidized-bed gasification. Biomass & Bioenergy. 2001;20(1):45-56. PubMed PMID: WOS:000167001400006.

[29] Gordillo G, Annamalai K. Adiabatic fixed bed gasification of dairy biomass with air and steam. Fuel. 2010 Feb;89(2):384-91. PubMed PMID: WOS:000271295400017.

[30] Boateng AA, Banowetz GM, Steiner JJ, Barton TF, Taylor DG, Hicks KB, et al. Gasification of Kentucky bluegrass (Poa pratensis l.) straw in a farm-scale reactor. Biomass & Bioenergy. 2007 Feb-Mar;31(2-3):153-61. PubMed PMID: WOS:000244170800006.

[31] Bridgwater T. Biomass for energy. Journal of the Science of Food and Agriculture. 2006 Sep;86(12):1755-68. PubMed PMID: WOS:000240412500004.

[32] Swanson RM, Platon A, Satrio JA, Brown RC. Techno-economic analysis of biomass-to-liquids production based on gasification. Fuel. 2010 Nov;89:S2-S10. PubMed PMID: WOS:000282368600002.

[33] Drake HL, Gössner AS, Daniel SL. Old acetogens, new light. Annals of the New York Academy of Sciences. 2008;1125:100-28.

[34] Thauer RK. Microbiology - A fifth pathway of carbon fixation. Science. 2007 Dec; 318(5857):1732-3. PubMed PMID: WOS:000251616800025.

[35] Drake HL, Küsel K, Matthies C. Acetogenic prokaryotes. In: Dworkin M, Falkow S, Rosenberg E, Schleifer K-H, Stackebrandt E, editors.: Springer New York; 2006. p. 354-420.

[36] Lindahl PA, Chang B. The evolution of acetyl-CoA synthase. Origins of Life and Evolution of the Biosphere. 2001;31(4-5):403-34. PubMed PMID: WOS:000170975700006.

[37] Ragsdale SW. Enzymology of the Wood-Ljungdahl pathway of acetogenesis. Incredible Anaerobes: from Physiology to Genomics to Fuels. 2008;1125:129-36. PubMed PMID: WOS:000255395600007.

[38] Ragsdale SW. Life with carbon monoxide. Critical Reviews in Biochemistry and Molecular Biology. 2004 May-Jun;39(3):165-95. PubMed PMID: WOS:000223849300003.

[39] Ragsdale SW, Clark JE, Ljungdahl LG, Lundie LL, Drake HL. Properties of purified carbon monoxide dehydrogenase from Clostridium thermoaceticum, a nickel, iron-sulfur protein. Journal of Biological Chemistry. 1983;258(4):2364-9. PubMed PMID: WOS:A1983QD12900054. English.

[40] Morsdorf G, Frunzke K, Gadkari D, Meyer O. Microbial growth oncarbon monoxide. Biodegradation. 1992;3:61-82.

[41] Genthner BRS, Bryant MP. Additional characteristics of one-carbon-compound utilization by Eubacterium limosum and Acetobacterium woodi. Applied and Environmental Microbiology. 1987 Mar;53(3):471-6. PubMed PMID: WOS:A1987G206800001.

[42] Poehlein A, Schmidt S, Kaster AK, Goenrich M, Vollmers J, Thurmer A, et al. An ancient pathway combining carbon dioxide fixation with the generation and utilization of a sodium ion gradient for ATP synthesis. Plos One. 2012 Mar;7(3). PubMed PMID: WOS:000304523400022.

[43] Kane MD, Breznak JA. Acetonema longum gen nov SP-nov, an H2/CO2 acetogenic bacterium from the termite, Pterotermes occidentis. Archives of Microbiology. 1991 Aug;156(2):91-8. PubMed PMID: WOS:A1991GB27400002.

[44] Allen TD, Caldwell ME, Lawson Pa, Huhnke RL, Tanner RS. Alkalibaculum bacchi gen. nov., sp. nov., a CO-oxidizing, ethanol-producing acetogen isolated from livestock-impacted soil. International journal of systematic and evolutionary microbiology. 2010;60(Pt 10):2483-9.

[45] Liu K, Atiyeh HK, Tanner RS, Wilkins MR, Huhnke RL. Fermentative production of ethanol from syngas using novel moderately alkaliphilic strains of Alkalibaculum bacchi. Bioresource Technology. 2012 Jan;104:336-41. PubMed PMID: WOS: 000301155800045.

[46] Lorowitz WH, Bryant MP. Peptostreptococcus productus strain that grows rapidly with CO as the energy source. Applied and Environmental Microbiology. 1984;47(5): 961-4. PubMed PMID: WOS:A1984SQ91300014.

[47] Lynd L, Kerby R, Zeikus JG. Carbon monoxide metabolism of the methylotrophic acidogen Butyribacterium methylotrophicum. Journal of bacteriology. 1982;149(1): 255-63.

[48] Zeikus JG, Lynd LH, Thompson TE, Krzycki Ja, Weimer PJ, Hegge PW. Isolation and characterization of a new, methylotrophic, acidogenic anaerobe, the marburg strain. Current Microbiology. 1980;3(6):381-6.

[49] Grethlein AJ, Worden RM, Jain MK, Datta R. Evidence for production of n-butanol from carbon monoxide by Butyribacterium methylotrophicum. Journal of Fermentation and Bioengineering. 1991;72(1):58-60.

[50] Lux MF, Drake HL. Reexamination of the metabolic potentials of the acetogens Clostridium aceticum and Clostridium formicoaceticum - chemolithoautotrophic and aromatic-dependent growth. Fems Microbiology Letters. 1992 Aug;95(1):49-56. PubMed PMID: WOS:A1992JJ52500008.

[51] Adamse AD. New isolation of Clostridium aceticum (Wieringa). Antonie Van Leeuwenhoek Journal of Microbiology. 1980;46(6):523-31. PubMed PMID: WOS:A1980LA02100002.

[52] Braun M, Mayer F, Gottschalk G. Clostridium aceticum (Wieringa), A microorganism producing acetic acid from molecular hydrogen and carbon dioxide. Archives of Microbiology. 1981;128(3):288-93. PubMed PMID: WOS:A1981LB03200006.

[53] Abrini J, Naveau H, Nyns EJ. Clostridium autoethanogenum, sp. nov., an anaerobic bacterium that produces ethanol from carbon monoxide. Archives of microbiology. 1994;161(4):345-51.

[54] Bruant G, Lévesque M-J, Peter C, Guiot SR, Masson L. Genomic analysis of carbon monoxide utilization and butanol production by Clostridium carboxidivorans strain P7. PloS one. 2010;5(9):e13033.

[55] Liou JSC, Balkwill DL, Drake GR, Tanner RS. Clostridium carboxidivorans sp nov., a solvent-producing clostridium isolated from an agricultural settling lagoon, and reclassification of the acetogen Clostridium scatologenes strain SL1 as Clostridium drakei sp nov. International Journal of Systematic and Evolutionary Microbiology. 2005 Sep;55:2085-91. PubMed PMID: WOS:000232239600055.

[56] Küsel K, Dorsch T, Acker G, Stackebrandt E, Drake HL. Clostridium scatologenes strain SL1 isolated as an acetogenic bacterium from acidic sediments. International journal of systematic and evolutionary microbiology. 2000;50 Pt 2:537-46.

[57] Gössner AS, Picardal F, Tanner RS, Drake HL. Carbon metabolism of the moderately acid-tolerant acetogen Clostridium drakei isolated from peat. FEMS microbiology letters. 2008;287(2):236-42.

[58] Andreese.Jr, Gottscha.G, Schlegel HG. Clostridium formicoaceticumnov spec - isolation, description and distinction from C aceticum and C thermoaceticum. Archiv Fur Mikrobiologie. 1970;72(2):154. PubMed PMID: WOS:A1970G704100007.

[59] Diekert GB, Thauer RK. Carbon monoxide oxidation by Clostridium thermoaceticum and Clostridium formicoaceticum. Journal of bacteriology. 1978;136(2):597-606.

[60] Kusel K, Karnholz A, Trinkwalter T, Devereux R, Acker G, Drake HL. Physiological ecology of Clostridium glycolicum RD-1, an aerotolerant acetogen isolated from sea grass roots. Applied and Environmental Microbiology. 2001 Oct;67(10):4734-41. PubMed PMID: WOS:000171237700045.

[61] Ohwaki K, Hungate RE. Hydrogen utilization by Clostridia in sewage sludge. Applied and Environmental Microbiology. 1977;33(6):1270-4.

[62] Köpke M, Held C, Hujer S, Liesegang H, Wiezer A, Wollherr A, et al. Clostridium ljungdahlii represents a microbial production platform based on syngas. Proceedings of the National Academy of Sciences of the United States of America. 2010;107(29): 13087-92.

[63] Tanner RS, Miller LM, Yang D. Clostridium ljungdahlii sp. nov., an acetogenic species in clostridial rRNA homology group I. International journal of systematic bacteriology. 1993;43(2):232-.

[64] Bomar M, Hippe H, Schink B. Lithotrophic growth and hydrogen metabolism by Clostridium magnum. Fems Microbiology Letters. 1991;83(3):347-50.

[65] Schink B. Clostridium magnum sp nov, a non-autotrophic homoacetogenic bacterium. Archives of Microbiology. 1984;137(3):250-5.

[66] Kane MD, Brauman A, Breznak JA. Clostridium mayombei sp nov, an H2/CO2 acetogenic bacterium from the gut of the african soil-feeding termite, Cubitermes speciosus. Archives of Microbiology. 1991;156(2):99-104.

[67] Mechichi T, Labat M, Patel BKC, Woo THS, Thomas P, Garcia JL. Clostridium methoxybenzovorans sp nov., a new aromatic o-demethylating homoacetogen from an olive mill wastewater treatment digester. International Journal of Systematic Bacteriology. 1999 Jul;49:1201-9. PubMed PMID: WOS:000081448000032.

[68] Huhnke R, Lewis RS, Tanner RS, inventors. Isolation and characterization of novel Clostridial species. 2008.

[69] Genthner BRS, Davis CL, Bryant MP. Features of rumen and sewage sludge strains of Eubacterium limosum, a methanol-utilizing and H2-CO2-utilizing species. Applied and Environmental Microbiology. 1981;42(1):12-9. PubMed PMID: WOS:A1981LX19300003.

[70] Krumholz LR, Bryant MP. Clostridium pfennigii sp nov uses methoxyl groups of monobenzenoids and produces butyrate. International Journal of Systematic Bacteriology. 1985;35(4):454-6. PubMed PMID: WOS:A1985ASV9200009.

[71] Pierce E, Xie G, Barabote RD, Saunders E, Han CS, Detter JC, et al. The complete genome sequence of Moorella thermoacetica (f. Clostridium thermoaceticum). Environmental microbiology. 2008;10(10):2550-73.

[72] Daniel SL, Hsu T, Dean SI, Drake HL. Characterization of the H2-dependent and CO-dependent chemolithotrophic potentials of the acetogens Clostridium thermoaceticum and Acetogenium kivui. Journal of Bacteriology. 1990 Aug;172(8):4464-71. PubMed PMID: WOS:A1990DR47100043.

[73] Kerby R, Zeikus JG. Growth of Clostridium thermoaceticum on H2/CO2 or CO as energy source. Current Microbiology. 1983;8(1):27-30. PubMed PMID: WOS:A1983QB38000006.

[74] Savage MD, Wu ZR, Daniel SL, Lundie LL, Drake HL. Carbon monoxide-dependent chemolithotrophic growth of Clostridium thermoautotrophicum. Applied and Environmental Microbiology. 1987 Aug;53(8):1902-6. PubMed PMID: WOS:A1987J392000033.

[75] Müller V. Energy conservation in acetogenic bacteria. Applied and environmental microbiology. 2003;69(11):6345-53.

[76] Calusinska M, Happe T, Joris B, Wilmotte A. The surprising diversity of clostridial hydrogenases: a comparative genomic perspective. Microbiology-Sgm. 2010 Jun; 156:1575-88. PubMed PMID: WOS:000279798900001.

[77] Vignais PM, Colbeau A. Molecular biology of microbial hydrogenases. Current Issues in Molecular Biology. 2004;6:159-88. PubMed PMID: WOS:000228691500013.

[78] Vignais PM. Hydrogenases and H^+-reduction in primary energy conservation. In: Schafer G, Penefsky HS, editors. Bioenergetics: Energy Conservation and Conversion. 45: Springer Berlin / Heidelberg; 2007. p. 223-52.

[79] Demuez M, Cournac L, Guerrini O, Soucaille P, Girbal L. Complete activity profile of Clostridium acetobutylicum FeFe -hydrogenase and kinetic parameters for endogenous redox partners. Fems Microbiology Letters. 2007 Oct;275(1):113-21. PubMed PMID: WOS:000249428700015.

[80] Schut GJ, Adams MWW. The iron-hydrogenase of Thermotoga maritima utilizes ferredoxin and NADH synergistically: a new perspective on anaerobic hydrogen production. Journal of bacteriology. 2009;191(13):4451-7.

[81] Menon S. Unleashing hydrogenase activity in carbon monoxide dehydrogenase/ acetyl-CoA synthase and pyruvate: ferredoxin oxidoreductase. Biochemistry. 1996;2960(96):15814-21.

[82] Thauer RK, Jungermann K, Decker K. Energy conservation in chemotrophic anerobic bacteria. Bacteriological Reviews. 1977;41(1):100-80. PubMed PMID: WOS:A1977DC99200003.

[83] Oelgeschläger E, Rother M. Carbon monoxide-dependent energy metabolism in anaerobic bacteria and archaea. Archives of microbiology. 2008;190(3):257-69.

[84] Rohde M, Mayer F, Meyer O. Immunocytochemical localization of carbon-monoxide oxidase in Pseudomonas carboxydovorans - The enzyme is attached to the inner aspect of the cytoplasmic membrane. Journal of Biological Chemistry. 1984;259(23): 4788-92. PubMed PMID: WOS:A1984TV66900080.

[85] Rohde M, Mayer F, Jacobitz S, Meyer O. Attachement of CO dehydrogenase to the cytoplasmic membrane is limiting the respiratory rate of Pseudomonas carboxydovorans. Fems Microbiology Letters. 1985;28(2):141-4. PubMed PMID: WOS:A1985AMV6400003.

[86] Kumar M, Lu WP, Ragsdale SW. Binding of carbon disulfide to the site of acetyl-CoA synthesis by the nickel-iron-sulfur protein, carbon monoxide dehydrogenase, from Clostridium thermoaceticum. Biochemistry. 1994 Aug;33(32):9769-77. PubMed PMID: WOS:A1994PC54500048.

[87] Lindahl PA, Munck E, Ragsdale SW. CO dehydrogenase from Clostridium thermoaceticum - EPR and electrochemical studies in CO2 and argon atmosphere. Journal of Biological Chemistry. 1990 Mar;265(7):3873-9. PubMed PMID: WOS:A1990CQ48000047.

[88] Ragsdale SW, Pierce E. Acetogenesis and the Wood-Ljungdahl pathway of CO(2) fixation. Biochimica et biophysica acta. 2008;1784(12):1873-98.

[89] Nevin KP, Hensley SA, Franks AE, Summers ZM, Ou JH, Woodard TL, et al. Electrosynthesis of organic compounds from carbon dioxide is catalyzed by a diversity of acetogenic microorganisms. Applied and Environmental Microbiology. 2011 May; 77(9):2882-6. PubMed PMID: WOS:000289773100009.

[90] Nevin KP, Woodard TL, Franks AE, Summers ZM, Lovley DR. Microbial electrosynthesis : Feeding microbes electricity to convert carbon dioxide and water to multicarbon extracellular organic compounds. mBio. 2010;1(2).

[91] Biegel E, Schmidt S, Müller V. Genetic, immunological and biochemical evidence for a Rnf complex in the acetogen Acetobacterium woodii. Environmental microbiology. 2009;11(6):1438-43.

[92] Li F, Hinderberger J, Seedorf H, Zhang J, Buckel W, Thauer RK. Coupled ferredoxin and crotonyl coenzyme A (CoA) reduction with NADH catalyzed by the butyryl-CoA dehydrogenase/Etf complex from Clostridium kluyveri. Journal of bacteriology. 2008;190(3):843-50.

[93] Müller V, Imkamp F, Biegel E, Schmidt S, Dilling S. Discovery of a ferredoxin:NAD+-oxidoreductase (Rnf) in Acetobacterium woodii: a novel potential coupling site in acetogens. Annals of the New York Academy of Sciences. 2008;1125:137-46.

[94] Herrmann G, Jayamani E, Mai G, Buckel W. Energy conservation via electron-transferring flavoprotein in anaerobic bacteria. Journal of Bacteriology. 2008 Feb;190(3): 784-91. PubMed PMID: WOS:000252616800002.

[95] Dürre P. Handbook on Clostridia. Dürre P, editor: CRC Press Taylor & Francis Group; 2005.

[96] Wang SN, Huang HY, Moll J, Thauer RK. NADP(+) reduction with reduced ferredoxin and NADP(+) reduction with NADH are coupled via an electron-bifurcating enzyme complex in Clostridium kluyveri. Journal of Bacteriology. 2010 Oct;192(19): 5115-23. PubMed PMID: WOS:000281866900032.

[97] Roberts DL, Jameshagstrom JE, Garvin DK, Gorst CM, Runquist JA, Baur JR, et al. Cloning and expression of the gene-cluster encoding key proteins involved in acetyl-CoA synthesis in Clostridium thermoaceticum - CO dehydrogenase, the corrinoid Fe-S protein, and methyltransferase. Proceedings of the National Academy of Sciences of the United States of America. 1989 Jan;86(1):32-6. PubMed PMID: WOS:A1989R820200008.

[98] Kaster AK, Goenrich M, Seedorf H, Liesegang H, Wollherr A, Gottschalk G, et al. More than 200 genes required for methane formation from H-2 and CO2 and energy conservation are present in Methanothermobacter marburgensis and Methanothermobacter thermautotrophicus. Archaea-an International Microbiological Journal. 2011. PubMed PMID: WOS:000292403100001.

[99] Papoutsakis ET, AL-Hinai MA, Jones SW, Indurthi DC, Mitchell DK, Fast A, inventors. Recombinant Clostridia that fix CO2 and CO and uses thereof. U.S.2012.

[100] Alsaker KV, Papoutsakis ET. Transcriptional program of early sporulation and stationary-phase events in Clostridium acetobutylicum. Journal of bacteriology. 2005;187(20):7103-18.

[101] Köpke M. Genetische Veränderung von Clostridium ljungdahlii zur produktion von 1-butanol aus synthesegas.: Ulm University; 2009.

[102] Sakai S, Nakashimada Y, Yoshimoto H, Watanabe S, Okada H, Nishio N. Ethanol production from H-2 and CO2 by a newly isolated thermophilic bacterium, Moorella sp HUC22-1. Biotechnology Letters. 2004 Oct;26(20):1607-12. PubMed PMID: WOS: 000225691400013.

[103] Sakai S, Nakashimada Y, Inokuma K, Kita M, Okada H, Nishio N. Acetate and ethanol production from H-2 and CO2 by Moorella sp using a repeated batch culture. Journal of Bioscience and Bioengineering. 2005 Mar;99(3):252-8. PubMed PMID: WOS:000229052300009.

[104] Köpke M, Mihalcea C, Liew FM, Tizard JH, Ali MS, Conolly JJ, et al. 2,3-butanediol production by acetogenic bacteria, an alternative route to chemical synthesis, using industrial waste gas. Applied and Environmental Microbiology. 2011 Aug;77(15): 5467-75. PubMed PMID: WOS:000293224500045.

[105] Köpke M, Mihalcea C, Bromley JC, Simpson SD. Fermentative production of ethanol from carbon monoxide. Current Opinion in Biotechnology. 2011 Jun;22(3):320-5. PubMed PMID: WOS:000292228300003.

[106] Kundiyana DK, Wilkins MR, Maddipati P, Huhnke RL. Effect of temperature, pH and buffer presence on ethanol production from synthesis gas by "Clostridium ragsdalei". Bioresource technology. 2011;102(10):5794-9.

[107] Saxena J, Tanner RS. Effect of trace metals on ethanol production from synthesis gas by the ethanologenic acetogen, Clostridium ragsdalei. Journal of Industrial Microbiology & Biotechnology. 2011 Apr;38(4):513-21. PubMed PMID: WOS: 000288673700004.

[108] Heiskanen H, Virkajärvi I, Viikari L. The effect of syngas composition on the growth and product formation of Butyribacterium methylotrophicum. Enzyme and Microbial Technology. 2007;41(3):362-7.

[109] Noelling J. Genome sequence and comparative analysis of the solvent-producing bacterium Clostridium acetobutylicum. Journal of. 2001;183(16):4823-38.

[110] Huang H, Liu H, Gan Y-R. Genetic modification of critical enzymes and involved genes in butanol biosynthesis from biomass. Biotechnology advances. 2010;28(5): 651-7.

[111] Ukpong MN, Atiyeh HK, De Lorme MJM, Liu K, Zhu X, Tanner RS, et al. Physiological response of Clostridium carboxidivorans during conversion of synthesis gas to

solvents in a gas-fed bioreactor. Biotechnology and bioengineering. 2012;Accepted article.

[112] Kundiyana DK, Huhnke RL, Wilkins MR. Syngas fermentation in a 100-L pilot scale fermentor: Design and process considerations. Journal of Bioscience and Bioengineering. 2010 May;109(5):492-8. PubMed PMID: WOS:000279089600011.

[113] Phillips JR, Klasson KT, Claussen EC, Gaddy JL. Biological production of ethanol from coal synthesis gas. Applied Biochemistry And Biotechnology. 1993;39(1):0-7.

[114] Lee SY, Park JH, Jang SH, Nielsen LK, Kim J, Jung KS. Fermentative butanol production by clostridia. Biotechnology and Bioengineering. 2008 Oct;101(2):209-28. PubMed PMID: WOS:000259512900001.

[115] Sauer U, Durre P. Differential induction of genes related to solvent formation during the shift from acidogenesis to solventogenesis in continuous culture of Clostridium acetobutylicum. Fems Microbiology Letters. 1995 Jan;125(1):115-20. PubMed PMID: WOS:A1995QA11200018.

[116] Shi Z, Blaschek HP. Transcriptional analysis of Clostridium beijerinckii NCIMB 8052 and the hyper-butanol-producing mutant BA101 during the shift from acidogenesis to solventogenesis. Applied and Environmental Microbiology. 2008 Dec;74(24): 7709-14. PubMed PMID: WOS:000261513700031.

[117] Köpke M, Dürre P. Biochemical production of biobutanol. In: Luque R, Campelo J, Clark JH, editors. Camebridge, UK: Woodhead Publishing Ltd; 2011. p. 221-57.

[118] Gheshlaghi R, Scharer JM, Moo-young M, Chou CP. Metabolic pathways of clostridia for producing butanol. Biotechnology advances. 2009;27(6):764-81.

[119] Seedorf H, Fricke WF, Veith B, Bruggemann H, Liesegang H, Strittimatter A, et al. The genome of Clostridium kluyveri, a strict anaerobe with unique metabolic features. Proceedings of the National Academy of Sciences of the United States of America. 2008 Feb;105(6):2128-33. PubMed PMID: WOS:000253261900064.

[120] Inui M, Suda M, Kimura S, Yasuda K, Suzuki H, Toda H, et al. Expression of Clostridium acetobutylicum butanol synthetic genes in Escherichia coli. Applied microbiology and biotechnology. 2008;77(6):1305-16.

[121] Tomas CA, Beamish J, Papoutsakis ET. Transcriptional analysis of butanol stress and tolerance in Clostridium acetobutylicum. Journal of Bacteriology. 2004 Apr;186(7): 2006-18. PubMed PMID: WOS:000220363200011.

[122] Jones DT, Woods DR. Acetone-butanol fermentation revisited. Microbiological reviews. 1986;50(4):484-524.

[123] Paul D, Austin FW, Arick T, Bridges SM, Burgess SC, Dandass YS, et al. Genome sequence of the solvent-producing bacterium Clostridium carboxidivorans strain P7T. Journal of bacteriology. 2010;192(20):5554-5.

[124] Ismaiel aa, Zhu CX, Colby GD, Chen JS. Purification and characterization of a primary-secondary alcohol dehydrogenase from two strains of Clostridium beijerinckii. Journal of bacteriology. 1993;175(16):5097-105.

[125] Ramachandriya KD, Wilkins MR, Delorme MJM, Zhu X, Kundiyana DK, Atiyeh HK, et al. Reduction of acetone to isopropanol using producer gas fermenting microbes. Biotechnology and bioengineering. 2011;108(10):2330-8.

[126] Toth J, Ismaiel aa, Chen JS. The ald gene, encoding a coenzyme A-acylating aldehyde dehydrogenase, distinguishes Clostridium beijerinckii and two other solvent-producing clostridia from Clostridium acetobutylicum. Applied and environmental microbiology. 1999;65(11):4973-80.

[127] Bertram J, Kuhn A, Durre P. Tn916-induced mutants of Clostridium acetobutylicum defective in regulation of solvent formation. Archives of Microbiology. 1990 Mar; 153(4):373-7. PubMed PMID: WOS:A1990CW24900012.

[128] Durre P, Fischer RJ, Kuhn A, Lorenz K, Schreiber W, Sturzenhofecker B, et al. Solventogenic enzymes of Clostridium acetobutylicum - catalytic properties, genetic organization, and transcriptional regulation. Fems Microbiology Reviews. 1995 Oct;17(3): 251-62. PubMed PMID: WOS:A1995TA64700005.

[129] Thauer RK, Jungerma.Ka, Kirchnia.Fh. Properties and function of pyruvate-formate-lyase reaction in Clostridiae. European Journal of Biochemistry. 1972;27(2):282-90. PubMed PMID: WOS:A1972M614500011.

[130] Hemme CL, Mouttaki H, Lee Y-J, Zhang G, Goodwin L, Lucas S, et al. Genome Announcement - Sequencing of multiple Clostridia genomes related to biomass conversion and biofuels production. Journal of bacteriology. 2010;192(24):6494-6.

[131] Wang Y, Li X, Mao Y, Blaschek HP. Genome-wide dynamic transcriptional profiling in Clostridium beijerinckii NCIMB 8052 using single-nucleotide resolution RNA-Seq. BMC genomics. 2012;13(1):102-.

[132] Janssen H, Döring C, Ehrenreich A, Voigt B, Hecker M, Bahl H, et al. A proteomic and transcriptional view of acidogenic and solventogenic steady-state cells of Clostridium acetobutylicum in a chemostat culture. Applied microbiology and biotechnology. 2010;87(6):2209-26.

[133] Haus S, Jabbari S, Millat T, Janssen H, Fischer R-J, Bahl H, et al. A systems biology approach to investigate the effect of pH-induced gene regulation on solvent production by Clostridium acetobutylicum in continuous culture. BMC systems biology. 2011;5(1):10-.

[134] Alsaker KV, Paredes C, Papoutsakis ET. Metabolite stress and tolerance in the production of biofuels and chemicals: gene-expression-based systems analysis of butanol, butyrate, and acetate stresses in the anaerobe Clostridium acetobutylicum. Biotechnology and bioengineering. 2010;105(6):1131-47.

[135] Allcock ER, Reid SJ, Jones DT, Woods DR. Autolytic activity of an autolysin-deficient mutant of Clostridium acetobutylicum. Appllied Environmental Microbiology. 1981;42:929-35.

[136] Hermann M, Fayolle F, Marchal R, Podvin L, Sebald M, Vandecasteele JP. Isolation and characterization of butanol-resistant mutants of Clostridium acetobutylicum. Applied Environmental Microbiology. 1985;50:1238-43.

[137] Vanderwesthuizen A, Jones DT, Woods DR. Autolytic activity and butanol tolerance of Clostridium acetobutylicum. Applied and Environmental Microbiology. 1982;44(6):1277-81. PubMed PMID: WOS:A1982PT27700005.

[138] Baer SH, Blaschek HP, Smith TL. Effect of butanol challenge and temperature on lipid composition and membrane fluidity of butanol-tolerant Clostridium acetobutylicum. Applied and environmental microbiology. 1987;53(12):2854-61.

[139] Soucaille P, Joliff G, Izard A, Goma G. Butanol tolerance and autobacteriocin production by Clostridium acetobutylicum. Current Microbiology. 1987;14(5):295-9. PubMed PMID: WOS:A1987G253100011.

[140] Mermelstein LD, Welker NE, Bennett GN, Papoutsakis ET. Expression of cloned homologous fermentative genes in Clostridium acetobutylicum ATCC 824. Bio-Technology. 1992 Feb;10(2):190-5. PubMed PMID: WOS:A1992HJ61200027.

[141] Mermelstein LD, Papoutsakis ET, Petersen DJ, Bennett GN. Metabolic engineering of Clostridium acetobutylicum ATCC 824 for increased solvent production by enhancement of acetone formation enzyme activities using a synthetic acetone operon. Biotechnology and Bioengineering. 1993 Nov;42(9):1053-60. PubMed PMID: WOS:A1993MC12100005.

[142] Walter KA, Mermelstein LD, Papoutsakis ET. Studies of recombinant Clostridium acetobutylicum with increased dosages of butyrate formation genes. Recombinant DNA Technology Ii. 1994;721:69-72. PubMed PMID: WOS:A1994BA43Z00006.

[143] Desai RP, Papoutsakis ET. Antisense RNA strategies for metabolic engineering of Clostridium acetobutylicum. Applied and Environmental Microbiology. 1999 Mar; 65(3):936-45. PubMed PMID: WOS:000078882000010.

[144] Sillers R, A-Hinai MA, Papoutsakis ET. Aldehyde-alcohol dehydrogenase and/or thiolase overexpression coupled with CoA transferase downregulation lead to higher alcohol titers and selectivity in Clostridium acetobutylicum fermentations. Biotechnology and Bioengineering. 2009 Jan 1;102(1):38-49. PubMed PMID: WOS: 000261826000006.

[145] Tummala SB, Junne SG, Papoutsakis ET. Antisense RNA downregulation of coenzyme A transferase combined with alcohol-aldehyde dehydrogenase overexpression leads to predominantly alcohologenic Clostridium acetobutylicum fermentations. Journal of Bacteriology. 2003 Jun;185(12):3644-53. PubMed PMID: WOS: 000183364700019.

[146] Green EM, Boynton ZL, Harris LM, Rudolph FB, Papoutsakis ET, Bennett GN. Genetic manipulation of acid formation pathways by gene inactivation in Clostridium acetobutylicum ATCC 824. Microbiology-Uk. 1996 Aug;142:2079-86. PubMed PMID: WOS:A1996VB82900020.

[147] Green EM, Bennett GN. Inactivation of an aldehyde/alcohol dehydrogenase gene from Clostridium acetobutylicum ATCC 824. Applied biochemistry and biotechnology. 1996;57-58:213-21.

[148] Wilkinson SR, Young M. Targeted integration of genes into the Clostridium acetobutylicum genome. Microbiology-Uk. 1994 Jan;140:89-95. PubMed PMID: WOS:A1994NE68400013.

[149] Shimizu T, Bathein W, Tamaki M, Hayashi H. The virr gene, a member of a class of 2-component response regulators, regulates the production of perfringolysin-o, collagenase, and hemagglutinin in Clostridium perfringens. Journal of Bacteriology. 1994 Mar;176(6):1616-23. PubMed PMID: WOS:A1994NB99400008.

[150] Liyanage H, Kashket S, Young M, Kashket ER. Clostridium beijerinckii and Clostridium difficile detoxify methylglyoxal by a novel mechanism involving glycerol dehydrogenase. Applied and Environmental Microbiology. 2001 May;67(5):2004-10. PubMed PMID: WOS:000168488400003.

[151] O'Connor JR, Lyras D, Farrow KA, Adams V, Powell DR, Hinds J, et al. Construction and analysis of chromosomal Clostridium difficile mutants. Molecular Microbiology. 2006 Sep;61(5):1335-51. PubMed PMID: WOS:000239701400020.

[152] Wilkinson SR, Young DI, Morris JG, Young M. Molecular genetics and the initiation of solventogenesis in Clostridium beijerinckii (formerly Clostridium acetobutylicum) NCIMB-8052. Fems Microbiology Reviews. 1995 Oct;17(3):275-85. PubMed PMID: WOS:A1995TA64700007.

[153] Harris LM, Welker NE, Papoutsakis ET. Northern, morphological, and fermentation analysis of spo0A inactivation and overexpression in Clostridium acetobutylicum ATCC 824. Journal of Bacteriology. 2002 Jul;184(13):3586-97. PubMed PMID: WOS: 000176237400020.

[154] Sarker MR, Carman RJ, McClane BA. Inactivation of the gene (cpe) encoding Clostridium perfringens enterotoxin eliminates the ability of two cpe-positive C-perfringens type A human gastrointestinal disease isolates to affect rabbit ileal loops. Molecular Microbiology. 1999 Sep;33(5):946-58. PubMed PMID: WOS:000082779700005.

[155] Soucaille P, Rainer F, Christian C, inventors. Process for chromosomal integration and DNA sequence replacement in clostridia. EU. 2008.

[156] Argyros DA, Tripathi SA, Barrett TF, Rogers SR, Feinberg LF, Olson DG, et al. High ethanol titers from cellulose by using metabolically engineered thermophilic, anaerobic microbes. Applied and Environmental Microbiology. 2011 Dec;77(23):8288-94. PubMed PMID: WOS:000297164100012.

[157] Karberg M, Guo HT, Zhong J, Coon R, Perutka J, Lambowitz AM. Group II introns as controllable gene targeting vectors for genetic manipulation of bacteria. Nature Biotechnology. 2001 Dec;19(12):1162-7. PubMed PMID: WOS:000172524400030.

[158] Heap JT, Pennington OJ, Cartman ST, Carter GP, Minton NP. The ClosTron: A universal gene knock-out system for the genus Clostridium. Journal of Microbiological Methods. 2007 Sep;70(3):452-64. PubMed PMID: WOS:000249573800008.

[159] Heap JT, Kuehne Sa, Ehsaan M, Cartman ST, Cooksley CM, Scott JC, et al. The ClosTron: Mutagenesis in Clostridium refined and streamlined. Journal of microbiological methods. 2010;80(1):49-55.

[160] Camiade E, Peltier J, Bourgeois I, Couture-Tosi E, Courtin P, Antunes A, et al. Characterization of Acp, a Peptidoglycan hydrolase of Clostridium perfringens with N-acetylglucosaminidase activity That is implicated in cell separation and stress-induced autolysis. Journal of Bacteriology. 2010 May;192(9):2373-84. PubMed PMID: WOS:000276685800009.

[161] Heap JT, Ehsaan M, Cooksley CM, Ng YK, Cartman ST, Winzer K, et al. Integration of DNA into bacterial chromosomes from plasmids without a counter-selection marker. Nucleic Acids Res. 2012 Apr 1;40(8):e59. PubMed PMID: 22259038. Pubmed Central PMCID: 3333862. Epub 2012/01/20. eng.

[162] Keis S, Shaheen R, Jones DT. Emended descriptions of Clostridium acetobutylicum and Clostridium beijerinckii, and descriptions of Clostridium saccharoperbutylacetonicum sp nov and Clostridium saccharobutylicum sp nov. International Journal of Systematic and Evolutionary Microbiology. 2001 Nov;51:2095-103. PubMed PMID: WOS:000172374400020.

[163] Bertram J, Durre P. Conjugal transfer and expression of Streptococcal transposons in Clostridium acetobutylicum. Archives of Microbiology. 1989 May;151(6):551-7. PubMed PMID: WOS:A1989AA14000015.

[164] Woolley RC, Pennock A, Ashton RJ, Davies A, Young M. Transfer of Tn1545 and Tn916 to Clostridium acetobutylicum. Plasmid. 1989 Sep;22(2):169-74. PubMed PMID: WOS:A1989CK56900010.

[165] Blouzard JC, Valette O, Tardif C, de Philip P. Random mutagenesis of Clostridium cellulolyticum by using a Tn1545 derivative. Applied and Environmental Microbiology. 2010 Jul;76(13):4546-9. PubMed PMID: WOS:000279082800049.

[166] Cartman ST, Minton NP. A mariner-based transposon system for in vivo random mutagenesis of Clostridium difficile. Appl Environ Microbiol. 2010 Feb;76(4):1103-9. PubMed PMID: 20023081. Pubmed Central PMCID: 2820977. Epub 2009/12/22. eng.

[167] Koepke M, Liew FM, inventors. Recombinant microorganism and methods of production thereof. U.S.2011.

[168] Simpson SD, Koepke M, Liew FM, inventors. Recombinant microorganisms and methods of use thereof. U.S.2011.

[169] Lederle S. Heterofermentative acetonproduktion: University of Ulm, Germany; 2011.

[170] Becker U, Grund G, Orschel M, Doderer K, Löhden G, Brand G, et al., inventors. Zellen und Verfahren zur Herstellung von Aceton. Germany. 2009.

[171] Harris LM, Desai RP, Welker NE, Papoutsakis ET. Characterization of recombinant strains of the Clostridium acetobutylicum butyrate kinase inactivation mutant: need for new phenomenological models for solventogenesis and butanol inhibition? Biotechnology and bioengineering. 2000;67(1):1-11.

[172] Lehmann D, Lutke-Eversloh T. Switching Clostridium acetobutylicum to an ethanol producer by disruption of the butyrate/butanol fermentative pathway. Metabolic Engineering. 2011 Sep;13(5):464-73. PubMed PMID: WOS:000294291200002.

[173] Kuit W, Minton NP, Lopez-Contreras AM, Eggink G. Disruption of the acetate kinase (ack) gene of Clostridium acetobutylicum results in delayed acetate production. Applied Microbiology and Biotechnology. 2012 May;94(3):729-41. PubMed PMID: WOS: 000302283700015.

[174] Yu MR, Zhang YL, Tang IC, Yang ST. Metabolic engineering of Clostridium tyrobutyricum for n-butanol production. Metabolic Engineering. 2011 Jul;13(4):373-82. PubMed PMID: WOS:000291471500003.

[175] Li YC, Tschaplinski TJ, Engle NL, Hamilton CY, Rodriguez M, Liao JC, et al. Combined inactivation of the Clostridium cellulolyticum lactate and malate dehydrogenase genes substantially increases ethanol yield from cellulose and switchgrass fermentations. Biotechnology for Biofuels. 2012 Jan;5. PubMed PMID: WOS: 000299821600001.

[176] Lee J, Jang YS, Choi SJ, Im JA, Song H, Cho JH, et al. Metabolic engineering of Clostridium acetobutylicum ATCC 824 for isopropanol-butanol-ethanol fermentation. Applied and Environmental Microbiology. 2012 Mar;78(5):1416-23. PubMed PMID: WOS:000300537400012.

[177] Higashide W, Li YC, Yang YF, Liao JC. Metabolic Engineering of Clostridium cellulolyticum for Production of isobutanol from cellulose. Applied and Environmental Microbiology. 2011 Apr;77(8):2727-33. PubMed PMID: WOS:000289459300020.

[178] Cai XP, Bennett GN. Improving the Clostridium acetobutylicum butanol fermentation by engineering the strain for co-production of riboflavin. Journal of Industrial Microbiology & Biotechnology. 2011 Aug;38(8):1013-25. PubMed PMID: WOS: 000293002000015.

[179] Zhu LJ, Dong HJ, Zhang YP, Li Y. Engineering the robustness of Clostridium acetobutylicum by introducing glutathione biosynthetic capability. Metabolic Engineering. 2011 Jul;13(4):426-34. PubMed PMID: WOS:000291471500008.

[180] Tomas CA, Welker NE, Papoutsakis ET. Overexpression of groESL in Clostridium acetobutylicum results in increased solvent production and tolerance, prolonged me-

tabolism, and changes in the cell's transcriptional program. Applied and Environmental Microbiology. 2003 Aug;69(8):4951-65. PubMed PMID: WOS: 000184672500078.

[181] Nakayama S-i, Kosaka T, Hirakawa H, Matsuura K, Yoshino S, Furukawa K. Metabolic engineering for solvent productivity by downregulation of the hydrogenase gene cluster hupCBA in Clostridium saccharoperbutylacetonicum strain N1-4. Applied Microbiology and Biotechnology. 2008 Mar;78(3):483-93. PubMed PMID: WOS: 000253132500012.

[182] Atsumi S, Hanai T, Liao J. Non-fermentative pathways for synthesis of branched-chain higher alcohols as biofuels. Nature. 2008;451(7174):86-9.

[183] Clomburg JM, Gonzalez R. Biofuel production in Escherichia coli: the role of metabolic engineering and synthetic biology. Applied Microbiology and Biotechnology. 2010 Mar;86(2):419-34. PubMed PMID: WOS:000274707600003.

[184] Peralta-Yahya PP, Keasling JD. Advanced biofuel production in microbes. Biotechnology Journal. 2010 Feb;5(2):147-62. PubMed PMID: WOS:000275275500005.

[185] Kapic A, Jones ST, Heindel TJ. Carbon monoxide mass transfer in a syngas mixture. Industrial & Engineering Chemistry Research. 2006 Dec;45(26):9150-5. PubMed PMID: WOS:000242786200049.

[186] Mohammadi M, Younesi H, Najafpour G, Mohamed AR. Sustainable ethanol fermentation from synthesis gas by Clostridium ljungdahlii in a continuous stirred tank bioreactor. Journal of Chemical Technology and Biotechnology. 2012 Jun;87(6): 837-43. PubMed PMID: WOS:000304190300018.

[187] Bredwell MD, Srivastava P, Worden RM. Reactor design issues for synthesis-gas fermentations. Biotechnology Progress. 1999 Sep-Oct;15(5):834-44. PubMed PMID: WOS:000082936200007.

[188] Ahmed A, Cateni BG, Huhnke RL, Lewis RS. Effects of biomass-generated producer gas constituents on cell growth, product distribution and hydrogenase activity of Clostridium carboxidivorans P7T. Biomass and Bioenergy. 2006;30(7):665-72.

[189] Xu D, Lewis RS. Syngas fermentation to biofuels: Effects of ammonia impurity in raw syngas on hydrogenase activity. Biomass and Bioenergy. 2012.

[190] Ahmed A, Lewis RS. Fermentation of biomass-generated synthesis gas: Effects of nitric oxide. Biotechnology and Bioengineering. 2007 Aug;97(5):1080-6. PubMed PMID: WOS:000247960300008.

[191] Brogren C, Karlsson HT, Bjerle I. Absorption of NO in an alkaline solution of KMnO4. Chemical Engineering & Technology. 1997 Aug;20(6):396-402. PubMed PMID: WOS:A1997XX28400006.

[192] Chu H, Chien TW, Li SY. Simultaneous absorption of SO2 and NO from flue gas with KMnO4/NaOH solutions. Science of the Total Environment. 2001 Jul;275(1-3): 127-35. PubMed PMID: WOS:000169969700011.

[193] Klasson KT, Ackerson MD, Clausen EC, Gaddy JL. Biological conversion of coal and coal-derived synthesis gas. Fuel. 1993 Dec;72(12):1673-8. PubMed PMID: WOS:A1993MJ23700016.

[194] Vega JL, Holmberg VL, Clausen EC, Gaddy JL. Fermentation parameters of Peptostreptococcus productus on gaseous substrates (CO, H2/CO2). Archives of Microbiology. 1989;151(1):65-70. PubMed PMID: WOS:A1989R169500012.

[195] Hurst KM, Lewis RS. Carbon monoxide partial pressure effects on the metabolic process of syngas fermentation. Biochemical Engineering Journal. 2010;48(2):159-65.

[196] Najafpour G, Younesi H. Ethanol and acetate synthesis from waste gas using batch culture of Clostridium ljungdahlii. Enzyme and Microbial Technology. 2006 Jan; 38(1-2):223-8. PubMed PMID: WOS:000234199300033.

[197] Gaddy JL, Chen GJ, inventors. Bioconversion of waste biomass to useful products. 1998.

[198] Genthner BRS, Bryant MP. Growth of Eubacterium limosum with carbon monoxide as the energy source. Applied and Environmental Microbiology. 1982;43(1):70-4. PubMed PMID: WOS:A1982MX20500009.

[199] Demler M, Weuster-Botz D. Reaction Engineering analysis of hydrogenotrophic production of ccetic ccid by Acetobacterium woodii. Biotechnology and Bioengineering. 2011 Feb;108(2):470-4. PubMed PMID: WOS:000285393000024.

[200] Cotter JL, Chinn MS, Grunden AM. Ethanol and acetate production by Clostridium ljungdahlii and Clostridium autoethanogenum using resting cells. Bioprocess Biosyst Eng. 2009;32(3):369-80.

[201] Kanchanatawee S, Maddox IS. Effect of phosphate and ammonium ion concentrations on solvent production in defined medium by Clostridium acetobutylicum. Journal of Industrial Microbiology. 1990 Jul;5(5):277-82. PubMed PMID: WOS:A1990DU68500002.

[202] Guo Y, Xu JL, Zhang Y, Xu HJ, Yuan ZH, Li D. Medium optimization for ethanol production with Clostridium autoethanogenum with carbon monoxide as sole carbon source. Bioresource Technology. 2010 Nov;101(22):8784-9. PubMed PMID: WOS: 000281262900044.

[203] Kundiyana DK, Huhnke RL, Maddipati P, Atiyeh HK, Wilkins MR. Feasibility of incorporating cotton seed extract in Clostridium strain P11 fermentation medium during synthesis gas fermentation. Bioresource Technology. 2010 Dec;101(24):9673-80. PubMed PMID: WOS:000282201200037.

[204] Sim JH, Kamaruddin AH. Optimization of acetic acid production from synthesis gas by chemolithotrophic bacterium - Clostridium aceticum using statistical approach.

Bioresource Technology. 2008 May;99(8):2724-35. PubMed PMID: WOS: 000254072200003.

[205] Vega JL, Prieto S, Elmore BB, Clausen EC, Gaddy JL. The biological production of ethanol from synthesis gas. Applied Biochemistry and Biotechnology. 1989 Jan-Aug; 20-1:781-97. PubMed PMID: WOS:A1989U642400058.

[206] Hu P, Jacobsen LT, Horton JG, Lewis RS. Sulfide assessment in bioreactors with gas replacement. Biochemical Engineering Journal. 2010 May;49(3):429-34. PubMed PMID: WOS:000277782900018.

[207] Do YS, Smeenk J, Broer KM, Kisting CJ, Brown R, Heindel TJ, et al. Growth of Rhodospirillum rubrum on synthesis gas: Conversion of CO to H-2 and poly-beta-hydroxyalkanoate. Biotechnology and Bioengineering. 2007 Jun;97(2):279-86. PubMed PMID: WOS:000246434700008.

[208] Devi MP, Mohan SV, Mohanakrishna G, Sarma PN. Regulatory influence of CO2 supplementation on fermentative hydrogen production process. International Journal of Hydrogen Energy. 2010 Oct;35(19):10701-9. PubMed PMID: WOS: 000283977100086.

[209] Grethlein AJ, Worden RM, Jain MK, Datta R. Continuous production of mixed alcohol and acids from carbon monoxide. Applied Biochemistry and Biotechnology. 1990 Spr-Sum;24-5:875-84. PubMed PMID: WOS:A1990DE57300079.

[210] Girbal L, Croux C, Vasconcelos I, Soucaille P. Regulation of metabolic shifts in Clostridium acetobutylicum ATCC 824. Fems Microbiology Reviews. 1995 Oct;17(3): 287-97. PubMed PMID: WOS:A1995TA64700008.

[211] Kundiyana DK, Wilkins MR, Maddipati P, Huhnke RL. Effect of temperature, pH and buffer presence on ethanol production from synthesis gas by "Clostridium ragsdalei". Bioresource Technology. 2011 May;102(10):5794-9. PubMed PMID: WOS: 000291125800033.

[212] Gaddy JL, Clausen EC, inventors. Clostridium ljungdahlii, an anaerobic ethanol and acetate producing microorganism. 1992.

[213] Cotter JL, Chinn MS, Grunden AM. Influence of process parameters on growth of Clostridium ljungdahlii and Clostridium autoethanogenum on synthesis gas. Enzyme and Microbial Technology. 2009 May;44(5):281-8. PubMed PMID: WOS: 000264908000006.

[214] Frohlich H, Villian L, Melzner D, Strube J. Membrane technology in bioprocess science. Chemie Ingenieur Technik. 2012 Jun;84(6):905-17. PubMed PMID: WOS: 000304438600024.

[215] Rushton A, Ward AS, Holdich RG. Solid-liquid filtration and separation technology. Germany: VCH Publishers, Inc.; 1996.

[216] Kargupta K, Datta S, Sanyal SK. Analysis of the performance of a continuous membrane bioreactor with cell recycling during ethanol fermentation. Biochemical Engineering Journal. 1998 Jan;1(1):31-7. PubMed PMID: WOS:000077894400004.

[217] Brown SD, Guss AM, Karpinets TV, Parks JM, Smolin N, Yang S, et al. Mutant alcohol dehydrogenase leads to improved ethanol tolerance in Clostridium thermocellum. Proceedings of the National Academy of Sciences of the United States of America. 2011;108(33):13752-7.

[218] Shao XJ, Raman B, Zhu MJ, Mielenz JR, Brown SD, Guss AM, et al. Mutant selection and phenotypic and genetic characterization of ethanol-tolerant strains of Clostridium thermocellum. Applied Microbiology and Biotechnology. 2011 Nov;92(3):641-52. PubMed PMID: WOS:000295673800020.

[219] Ezeji T, Milne C, Price ND, Blaschek HP. Achievements and perspectives to overcome the poor solvent resistance in acetone and butanol-producing microorganisms. Applied Microbiology and Biotechnology. 2010 Feb;85(6):1697-712. PubMed PMID: WOS:000273978400007.

[220] Borden JR, Papoutsakis ET. Dynamics of genomic-library enrichment and identification of solvent tolerance genes for Clostridium acetobutylicum. Applied and environmental microbiology. 2007;73(9):3061-8.

[221] Ezeji TC, Li Y. Advanced product recovery technologies. Hoboken: Wiley; 2009.

[222] Ezeji TC, Qureshi N, Blaschek HP. Bioproduction of butanol from biomass: from genes to bioreactors. Current opinion in biotechnology. 2007;18(3):220-7.

[223] Jauregui-Haza UJ, Pardillo-Fontdevila EJ, Wilhelm AM, Delmas H. Solubility of hydrogen and carbon monoxide in water and some organic solvents. Latin American Applied Research. 2004 Apr;34(2):71-4. PubMed PMID: WOS:000220679200001.

[224] Karcher P, Ezeji TC, Qureshi N, Blaschek HP. Microbial production of butanol: product recovery by extraction. New Delhi: IK International Publishing House Pvt Ltd; 2005.

[225] Ezeji TC, Li Y. Advanced product recovery technologies. Hoboken: Wiley; 2009.

[226] Ezeji TC, Qureshi N, Blaschek HP. Production of acetone, butanol and ethanol by Clostridium beijerinckii BA101 and in situ recovery by gas stripping. World Journal of Microbiology & Biotechnology. 2003 Aug;19(6):595-603. PubMed PMID: WOS: 000184717400007.

[227] Wei L, Pordesimo LO, Igathinathane C, Batchelor WD. Process engineering evaluation of ethanol production from wood through bioprocessing and chemical catalysis. Biomass & Bioenergy. 2009 Feb;33(2):255-66. PubMed PMID: WOS:000263995200011.

[228] Piccolo C, Bezzo F. A techno-economic comparison between two technologies for bioethanol production from lignocellulose. Biomass & Bioenergy. 2009 Mar;33(3): 478-91. PubMed PMID: WOS:000264433600016.

[229] Yoneda N, Kusano S, Yasui M, Pujado P, Wilcher S. Recent advances in processes and catalysts for the production of acetic acid. Applied Catalysis a-General. 2001 Nov;221(1-2):253-65. PubMed PMID: WOS:000173000700019.

[230] Zigova J, Sturdik E, Vandak D, Schlosser S. Butyric acid production by Clostridium butyricum with integrated extraction and pertraction. Process Biochemistry. 1999 Oct;34(8):835-43. PubMed PMID: WOS:000083166300009.

[231] Gaddy JL, inventor. Biological production of ethanol from waste gases with Clostridium ljungdahlii. US. 2000.

[232] Guzman DD. INEOS Bio looks to start US cellulosic ethanol in 2012. ICIS Chemical Business. 2011:22-3.

[233] Bio I. INEOS Bio - Pioneering waste-to-biofuel technology attracts cross-party support in UK 2011. Available from: http://www.ineosbio.com/76-Press_releases-16.htm.

[234] Coskata. Coskata, Inc.'s semi-commercial facility demonstrates two years of successful operation 2011. Available from: http://www.coskata.com/company/media.asp?story=504B571C-0916-474E-BFFA-ACB326EFDB68.

[235] Coskata. About us 2012. Available from: http://www.coskata.com/company/.

[236] Huhnke RL, Lewis RS, Tanner RS, inventors. Isolation and characterization of novel Clostridial species. US. 2010.

[237] Kundiyana DK, Huhnke RL, Wilkins MR. Effect of nutrient limitation and two-stage continuous fermentor design on productivities during "Clostridium ragsdalei" syngas fermentation. Bioresource Technology. 2011 May;102(10):6058-64. PubMed PMID: WOS:000291125800069.

[238] Zahn JA, Saxena J, inventors. Novel ethanologenic Clostridium species, Clostridium coskatii. US. 2012.

[239] Kuo I. Coskata grabs biggest slice of USDA's $405M in biofuels loan guarantees: Reuters; 2011. Available from: http://www.reuters.com/article/2011/01/21/idUS126007139820110121.

[240] Lane J. Coskata switches focus from biomass to natural gas; to raise $100M in natgas-oriented private placement: Biofuels Digest; 2012. Available from: http://www.biofuelsdigest.com/bdigest/2012/07/20/coskata-switches-from-biomass-to-natural-gas-to-raise-100m-in-natgas-oriented-private-placement/.

[241] Wernau J. Biofuel firm Coskata drops initial public offering plans amid volatile IPO market: Chicago Tribute Business; 2012. Available from: http://www.chicagotribune.com/business/ct-biz-0721-ipos-20120721,0,1092013.story.

[242] Lane J. The intrepid investor: How are biofuels IPOs performing and why? : Biofuels Digest; 2012. Available from: http://www.biofuelsdigest.com/bdigest/2012/02/09/the-intrepid-investor-part-i/.

[243] Guzman DD, Chang J. Bio-based 2,3 BDO set for 2014 sales. ICIS Chemical Business. 2012:24.

[244] LanzaTech. Media Releases. Available from: http://lanzatech.com/media/media-releases.

Chapter 4

Conversion of Oil Palm Empty Fruit Bunch to Biofuels

Anli Geng

Additional information is available at the end of the chapter

1. Introduction

Crude palm oil production is reaching 48.99 million metric tonnes per year globally in 2011 and Southeast Asia is the main contributor, with Indonesia accounting for 48.79%, Malaysia 36.75%, and Thailand 2.96% (Palm Oil Refiners Association of Malaysia, 2011). Oil palm is a multi-purpose plantation and it is also an intensive producer of biomass. Accompanying the production of one kg of palm oil, approximately 4 kg of dry biomass are produced. One third of the oil palm biomass is oil palm empty fruit bunch (OPEFB) and the other two thirds are oil palm trunks and fronds [1-3].

Figure 1. Oil palm and oil palm empty fruit bunch.

The supply of oil palm biomass and its processing by-products are found to be seven times that of natural timber [4]. Besides producing oils and fats, there are continuous interests in using oil palm biomass as the source of renewable energy. Among the oil palm biomass, OPEFB is the most often investigated biomass for biofuel production. Traditionally, OPEFB is used for power and steam utilization in the palm oil mills, and is used for composting and soil mulch. Direct burning of OPEFB causes environmental problems due the incomplete combustion and the release of very fine particles of ash. The conversion of OPEFB to biofuels, such as syngas, ethanol, butanol, bio-oil, hydrogen and biogas etc., might be a good alternative and have less environmental footprint. The properties of OPEFB is listed in Table 1 [5].

	Literature values % (w/w)	Measured % (w/w)	Method
Components			
Cellulose	59.7	na	na
Hemicellulose	22.1	na	na
Lignin	18.1	na	na
Eelemental analysis			
Carbon	48.9	49.07	Combustion analysis
Hydrogen	6.3	6.48	
Nitrogen	0.7	0.7	
Sulphur	0.2	<0.10	
Oxygen	36.7	38.29	By difference
K	2.24	2.00	Spectrometry
K_2O	3.08–3.65	na	na
Proximate analysis			
Moisture	na	7.95	ASTM E871
Volatiles	75.7	83.86	ASTM E872
Ash	4.3	5.36	NREL LAP005
Fixed carbon	17	10.78	By difference
HHV (MJ/kg)	19.0	19.35	Bomb calorimeter
LHV (MJ/kg)	17.2	na	na

Notes: na - not available.

Table 1. Properties of oil palm empty fruit bunch

While all the OPEFB components can be converted to biofuels, such as bio-oil and syngas through thermo-chemical conversion, cellulose and hemicellulose can be hydrolysed to sugars and subsequently be fermented to biofuels such as ethanol, butanol, and biogas etc. Although many scientists around the world are developing technologies to generate biofuels from OPEFB, to-date, none of such technologies has been commercialized. This is largely due to the recalcitrance of the OPEFB and therefore the complexity of the conversion technologies making biofuels from OPEFB less competitive than the fossil-based fuels. Continual efforts in R&D are still necessary in order to bring such technology to commercialization. The aim of this paper is to review the progress and challenges of the OPEFB conversion technologies so as to help expedite the OPEFB conversion technology development.

2. Pretreatment

Similar to all other lignocellulosic biomass, OPEFB are composed of cellulose, hemicellulose and lignin. Among the three components, lignin has the most complex structure, making it recalcitrant to both chemical and biological conversion. Pretreatment of OPEFB is therefore necessary to open its structure and increase its digestibility and subsequently the degree of conversion. Pretreatment of OPEFB can be classified as biological pretreatment, physical pretreatment, chemical pretreatment, and physical-chemical pretreatment.

For biological pretreatment, oxidizing enzymes and white-rot fungi were used to degrade the lignin content in OPEFB. For example, enzymes such as lignin peroxidase (LiP) and manganese peroxidase (MnP) was used to pretreat OPEFB for fast pyrolysis and the bio-oil yield was improved from 20% to 30% [6]. Syafwina et al. used white-rot fungi to pretreat OPEFB and the saccharification efficiency was improved by 150% compared to that of the untreated OPEFB [7].

Among all the pretreatment methods, chemical pretreatment is most often reported for OPEFB. Two-stage dilute acid hydrolysis [8], alkali pretreatment [9], sequential dilute acid and alkali pretreatment [10], alkali and hydrogen peroxide pretreatment [11], sequential alkali and phosphoric acid pretreatment [10], aqueous ammonia [12], and solvent digestion [5] were used to increase the digestibility of OPEFB. Among all the chemical methods investigated, alkali pretreatment seemed to be the most effective. Umikalsom et al. autoclaved the milled OPEFB in the presence of 2% NaOH and 85% hydrolysis yield was obtained [13]. Han and his colleagues investigated NaOH pretreatment of OPEFB for bioethanol production [9]. The optimal conditions were found to be 127.64°C, 22.08 min, and 2.89 mol/L NaOH. With a cellulase loading of 50 FPU /g cellulose a total glucose conversion rate (TGCR) of 86.37% was obtained using the Changhae Ethanol Multi Explosion (CHEMEX) facility. The effectiveness of alkali pretreatment might be attributed to its capability in lignin degradation. Mission et al. investigated the alkali treatment followed H_2O_2 treatment and found that almost 100% lignin degradation was obtained when OPEFB was firstly treated with dilute NaOH and subsequently with H_2O_2 [11]. This confirmed the lignin degradation by NaOH and its enhancement by the addition of H_2O_2.

Besides alkali pretreatment, physical-chemical pretreatment such as ammonium fibre explosion (AFEX) [14] and superheated steam [15] were also shown to be effective in the increase of OPEFB digestibility. Hydrolysis efficiency of 90% and 66% were obtained, respectively.

3. Thermo-chemical conversion

Thermo-chemical conversion is one of the important routes to obtain fuels from lignocellulosic biomass. Thermo-chemical conversion of biomass involves heating the biomass materials in the absence of oxygen to produce a mixture of gas, liquid and solid. Such products can be used as fuels after further conversion or upgrading. Generally, thermo-chemical processes have lower reaction time required (a few seconds or minutes) and the superior ability to destroy most of the organic compounds. These mainly include biomass pyrolysis and biomass gasification. Recently, thermo-chemical pretreatment of biomass, such as torrefaction was introduced to upgrade biomass for more efficient biofuel production [16-17].

3.1. OPEFB pyrolysis

Pyrolysis is defined as the thermal degradation of the biomass materials in the absence of oxygen. It is normally conducted at moderate temperature (400 – 600°C) over a short period of retention time. Its products comprise of liquids (water, oil/tars), solids (charcoal) and gases (methane, hydrogen, carbon monoxide and carbon dioxide). The efficiency of pyrolysis and the amount of solid, liquid, and gaseous fractions formed largely depend on the process parameters such as pretreatment condition, temperature, retention time and type of reactors.

Misson et al investigated the effects of alkaline pretreatment using NaOH, $Ca(OH)_2$, in conjunction with H_2O_2, on the catalytic pyrolysis of OPEFB [11]. They proved that consecutive addition of NaOH and H_2O_2 decomposed almost 100% of OPEFB lignin compared to 44% for the $Ca(OH)_2$ and H_2O_2 system, while the exclusive use of NaOH and $Ca(OH)_2$ could not alter lignin composition much. In addition, the pretreated OPEFB was catalytically pyrolysed more efficiently than the untreated OPEFB samples under the same conditions.

Fast pyrolysis represents a potential route to upgrade the OPEFB waste to value-added fuels and renewable chemicals. For woody feedstock, temperatures around 400-600°C together with short vapour residence times (0.5-2 s) are used to obtain bio-oil yields of around 70%, along with char and gas yields of around 15% each. Sulaiman and Abdullah investigated fast pyrolysis of OPEFB using and bench top fluidized bed reactor with a nominal capacity of 150 g/L [18]. After extensive feeding trials, it was found that only particles between 250 and 355 ˙m were easily fed. The maximum liquid and organics yields (55% total liquids) were obtained at 450°C. Higher temperature was more favourable for gas production and water content was almost constant in the range of temperature investigated. The maximum liquids yield and the minimum char yield were obtained at a residence time of 1.03 s. The pyrolysis liquids produced separated into two phases; a phase predominated by tarry organic compounds (60%) and an aqueous phase (40%). The phase separated liquid product would represent a challenging fuel for boilers and engines, due to the high viscosity of the organics phase and the high water content of the aqueous phase. These could be overcome by upgrading. However, the by-product, charcoal, has been commercialized for quite some time. It is worth noting that the first pilot bio-oil plant by Genting Bio-oil has already started operation in Malaysia [19].

3.2. OPEFB gasification

Gasification process is an extension of the pyrolysis process except that it is conducted at elevated temperature range of 800–1300 °C so that it is more favourable for gas production [20]. The gas stream is mainly composed of methane, hydrogen, carbon monoxide, and carbon dioxide. Biomass gasification offers several advantages, such as reduced CO_2 emissions, compact equipment requirements with a relatively small footprint, accurate combustion control, and high thermal efficiency. The main challenge in gasification is enabling the pyrolysis and gas reforming reactions to take place using the minimum amount of energy and gasifier design is therefore important [21].

Ogi et al. used an entrained-flow gasifier for OPEFB gasification at 900°C [22]. During gasification with H_2O alone, the carbon conversion rate was greater than 95% (C-equivalent), and hydrogen-rich gas with a composition suitable for liquid fuel synthesis ($[H_2]/[CO]$ = 1.8–3.9) was obtained. The gasification rate was improved to be greater than 99% when O_2 was added to H_2O; however, under these conditions, the gas composition was less suitable for liquid fuel synthesis due to the increase of CO_2 amount. Thermogravimetric (TG) analysis suggested that OPEFB decomposed easily, especially in the presence of H_2O and/or O_2, suggesting that OPEFB is an ideal candidate for biomass gasification. Lahijani and Zainal investigated OPEFB gasification in a pilot-scale air-blown fluidized bed reactor [23]. The effect of bed temperature (650–1050°C) on gasification performance was studied and the gasification results were compared to that of sawdust. Results showed that at 1050°C, OPEFB had almost equivalent gas yield and cold gas efficiency compared with saw dust, however, with low maximal heating values and higher carbon conversion. In addition, it was realized that agglomeration was the major issue in OPEFB gasification at high temperatures. This can be overcome by lowering the temperature to 770 ± 20 °C. Mohammed et al. studied OPEFB gasification in a bench scale fluidized-bed reactor for hydrogen-rich gas production [24]. The total gas yield was enhanced greatly with the increase of temperature and it reached the maximum value (~92 wt.%) at 1000 °C with big portions of H_2 (38.02 vol.%) and CO (36.36 vol.%). The feedstock particle size of 0.3–0.5 mm, was found to obtain a higher H_2 yield (33.93 vol.%), and higher LHV of gas product (15.26 MJ/m^3). The optimum equivalence ratio (ER) (0.25) was found to attain a higher H_2 yield (27.31 vol.%) at 850 °C. Due to the low efficiency of bench scale gasification unit the system needs to be scaling-up. The cost analysis for scale-up EFB gasification unit showed that the hydrogen supply cost is $2.11/kg OPEFB. Recently, a characterization and kinetic analysis was done by Mohammed et al. and it was found that a high content of volatiles (>82%) increased the reactivity of OPEFB, and more than 90% decomposed at 700 °C; however, a high content of moisture (>50%) and oxygen (>45%) resulted in a low calorific value [25]. The fuel characteristics of OPEFB are comparable to those of other biomasses and it can be considered a good candidate for gasification.

3.3. OPEFB torrefaction

Torrefaction is a thermal conversion method of biomass in the low temperature range of 200-300 °C. Biomass is pretreated to produce a high quality solid biofuel that can be used for combustion and gasification [16-17]. It is based on the removal of oxygen from biomass to produce a fuel with increased energy density. Different reaction conditions (temperature, inert gas, reaction time) and biomass resources lead to the differences in solid, liquid and gaseous products.

Uemura et al. [16] studied the effect of torrefaction on the basic characteristics of oil palm empty fruit bunches (EFB), mesocarp fibre and kernel shell as a potential source of solid fuel. It was found that mesocarp fibre and kernel shell exhibited excellent energy yield values higher than 95%, whereas OPEFB, on the other hand, exhibited a rather poor yield of 56%. Torrefaction can also be done in the presence of oxygen. Uemura and his colleagues [17] carried out OPEFB torrefaction in a fixed-bed tubular reactor in the presence of oxygen at varied oxygen concentration. The mass yield decreased with increasing temperature and oxygen concentration, but was unaffected by biomass particle size. The energy yield decreased with increasing oxygen concentrations, however, was still between 85% and 95%. It was found that the oxidative torrefaction process occurred in two successive steps or via two parallel reactions, where one reaction is ordinary torrefaction, and the other is oxidation.

3.4. Summary

The analysis of thermo-chemical conversion of OPEFB suggests that gasification is the most suitable thermo-chemical route for OPEFB conversion to biofuels. It has the highest carbon conversion (>90%) and biofuel yield. Due to the high viscosity and high water content of pyrolysis products, application of bio-oil as a biofuel is still very challenging. Compared to other oil palm residues, such as oil palm kernel, due to its high water content, OPEFB may not be a good candidate for solid fuels even after torrefaction pretreatment.

4. Bioconversion

Bioconversion of lignocellulosic biomass to fuels involves three major steps: 1) pretreatment-to effectively broken the biomass structure and release the biomass components i.e. cellulose, hemicellulose, and lignin, and therefore increase the digestibility of the biomass; 2) enzymatic hydrolysis – to hydrolyse cellulose and hemicellulose and produce fermentable sugar, such as glucose, xylose etc.; 3) fermentation – to convert the biomass hydrolysate sugars to the desired products. OPEFB was intensively investigated as a potential substrate for the production of biofuels, such as ethanol, butanol, and biogas etc. Among the biofuels produced through bioconversion of OPEFB, cellulosic ethanol is the most intensively studied.

Two stage dilute acid hydrolysis was applied for OPEFB bioconversion to ethanol, 135.94 g xylose/kg OPEFB and 62.70 g glucose/kg OPEFB were produced in the first stage and 2nd stage, respectively [8]. They were then fermented to ethanol using *Mucor indicus* and *Saccharomyces cerevisiae*, respectively, and the corresponding ethanol yields were 0.45 and 0.46 g ethanol/g sugar.

Alkali is the most often used pretreatment chemical for cellulosic ethanol production from OPEFB. Kassim et al. pretreated OPEFB using 1% NaOH followed by mild acid (0.7% H_2SO_4) hydrolysis and enzymatic saccharification [26]. A total of 16.4 g/L of glucose and 3.85 g/L of xylose were obtained during enzymatic saccharification. The OPEFB hydrolysate was fermented with *Saccharomyces cerevisiae* and an ethanol yield of 0.51 g/g yield was obtained, suggesting that OPEFB is a potential substrate for cellulosic ethanol production. Han and his colleagues investigated ethanol production through pilot scale alkali pretreatment and fermentation [9]. The best pretreatment condition was 127.64 °C, 22.08 min, and 2.89 mol/L

NaOH. Enzyme loading of 50 FPU/g cellulose resulted in 86.37% glucose conversion in their Changhae Ethanol Multi Explosion (CHEMEX) facility. An ethanol concentration of 48.54 g/L was obtained at 20% (w/v) pretreated biomass loading, along with simultaneous saccharification and fermentation (SSF) processes. This is so far the highest reported ethanol titre from OPEFB. Overall, 410.48 g of ethanol were produced from 3 kg of raw OPEFB in a single run, using the CHEMEX 50 L reactor.

Jung and his colleagues tried aqueous ammonia soaking for the pretreatment of OPEFB and its conversion to ethanol [12]. Pretreated OPEFB at 60°C, 12 h, and 21% (w/w) aqueous ammonia, showed 19.5% and 41.4% glucose yields after 96h enzymatic hydrolysis using 15 and 60 FPU of cellulase per gram of OPEFB, respectively. An ethanol concentration of 18.6 g/L and a productivity of 0.11 g/L/h were obtained with the ethanol yield of 0.33 g ethanol/ glucose.

Lau et al. successfully applied ammonia fibre expansion (AFEX) pretreatment for cellulosic ethanol production from OPEFB [14]. The sugar yield was close to 90% after enzyme formulation optimization. Post-AFEX size reduction is required to enhance the sugar yield possibly due to the high tensile strength (248 MPa) and toughness (2,000 MPa) of palm fibre compared to most cellulosic feedstock. Interestingly, the water extract from AFEX-pretreated OPEFB at 9% solids loading is highly fermentable and up to 65 g/L glucose can be fermented to ethanol within 24 h without the supplement of nutrients.

OPEFB was also used for butanol production. Noomtim and Cheirsilp (2011) studied butanol production from OPEFB using *Clostridium acetobutylicum* [27]. Again, the pretreatment by alkali was found to be the most suitable method to prepare OPEFB for enzymatic hydrolysis. 1.262 g/L ABE (acetone, butanol and ethanol) was obtained in RCM medium containing 20 g/L sugar obtained from cellulase hydrolysed OPEFB. Ibrahim et al also investigated OPEFB as the potential substrate for ABE production [28]. Higher ABE yield was obtained from treated OPEFB when compared to using a glucose-based medium using *Clostridium butyricum* EB6. A higher ABE level was obtained at pH 6.0 with a concentration of 3.47 g/L. The accumulated acid (5 to 13 g/L) had inhibitory effects on cell growth.

Nieves et al. investigated biogas production using OPEFB. OPEFB was pre-treated using NaOH and phosphoric acid [29]. When 8% NaOH (60 min) was used for the pretreatment, 100% improvement in the yield of methane production was observed and 97% of the theoretical value of methane production was achieved under such pretreatment condition. The results showed that the carbohydrate content of OPEFB could be efficiently converted to methane under the anaerobic digestion process. O-Thong et al. investigated the effect of pretreatment methods for improved biodegradability and biogas production of oil palm empty fruit bunches (EFB) and its co-digestion with palm oil mill effluent (POME) [30]. The maximum methane potential of OPEFB was 202 mL CH_4/g VS-added corresponding to 79.1 m^3 CH_4/ton OPEFB with 38% biodegradability. Co-digestion of treated OPEFB by NaOH presoaking and hydrothermal treatment with POME resulted in 98% improvement in methane yield comparing with co-digesting untreated OPEFB. The maximum methane production of co-digestion treated OPEFB with POME was 82.7 m^3 CH_4/ton of mixed treated OPEFB and POME (6.8:1), corresponding to methane yield of 392 mL CH_4/g VS-added. The study showed that there was a great potential to co-digestion treated OPEFB with POME for bio-energy production.

In summary, OPEFB has been frequently investigated as a substrate for biofuel production through bioconversion. Cellulosic ethanol production was most intensively investigated and the highest ethanol titre of 48.54 g/L was obtained through alkali pretreatment in a pilot scale reactor [9]. Although not much research has been done for ABE and biogas production, the few reports summarized in this paper suggest that OPEFB is also potential substrate for butanol and biogas production. Throughout the reports reviewed, alkali-based pretreatment methods, such as NaOH alone, NaOH followed by acid, and ammonium fibre expansion (AFEX) pretreatment are the most effective in enhancing OPEFB digestibility.

5. Conclusion

In conclusion, OPEFB is the most potential renewable resource for biofuel production in Southeast Asia. It can be converted to biofuels through thermo-chemical or biological conversion. Pretreatment of OPEFB is necessary for both routes of conversion and alkali pretreatment is the most effective. A summary of OPEFB conversion technology is shown in Fig. 2.

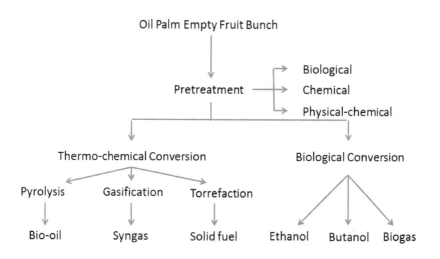

Figure 2. Biofuel production from OPEFB.

Among the studies on OPEFB thermo-chemical conversion, it seems that gasification is the most suitable approach to obtain bioenergy from OPEFB and has potential in commercialization. Pyrolysis, on the other hand, produced very complex bio-oil with high viscosity and water content, making it challenging for commercialization. However, charcoal from OPEFB pyrolysis can be a potential commercial product. Compared to other palm oil residues, such as oil palm kernel, OPEFB may not be a good candidate for solid fuel production, even after torrefaction pretreatment due to its high water content and low energy capacity.

Biological conversion of OPEFB is another route to obtain biofuels from OPEFB. Cellulosic ethanol production was most intensively studied and around 50 g/L titre was obtained with 20% (w/v) biomass loading through NaOH pretreatment. AFEX also showed potential in OPEFB pretreatment and a glucose yield of 90% was obtained with 9% biomass loading. The water extract of the AFEX pretreated OPEFB was highly fermentable. OPEFB also showed some promising preliminary results in ABE (acetone, butanol and ethanol) and biogas production; however, further investigation is necessary to enhance OPEFB conversion potentials in these areas.

For both thermo-chemical and biological conversion of OPEFB, pretreatment technology is the key for the process cost. Although alkali pretreatment is effective, scaling-up the process requires huge amount of acid to neutralize the base in the pretreatment solution. In addition, before alkali pretreatment, OPEFB should be milled to reduce its size, which is energy-consuming. Steam explosion is effective for a lot of lignocellulosic biomass, however not much research was found on its pretreatment of OPEFB. A cost-effective pretreatment is the key for the successful commercialization of OPEFB conversion technologies for biofuel production.

Acknowledgement

The. authors are grateful for the financial support to the research on cellulosic ethanol by the Science and Engineering Research Council of the Agency for Science, Technology and Research (A*STAR), Singapore.

Author details

Anli Geng

Address all correspondence to: gan2@np.edu.sg

School of Life Sciences and Chemical Technology, Ngee Ann Polytechnic, Singapore

References

[1] Husin M, Ramli R, Mokhtar A, Hassan WHW, Hassan K, Mamat R. Research and development of oil palm biomass utilization in wood-based industries. Palm Oil Development 2002; 36: 1-5.

[2]　Wahid MB, Abdullah SNA, Henson IE. Oil palm achievements and potential. New Directions for a diverse planet. In: Proceedings of the fourth international crop science congress, 26 Sep - 1 Oct 2004. Brisbane, Australia.

[3]　Yusoff S. Renewable energy from palm oil e innovation on effective utilization of waste. Journal of Cleaner Production 2006; 14: 87-93.

[4]　Basiron Y, Chan KW. The oil palm and its sustainability. Journal of Oil Palm Research 2004; 16: 1-10.

[5]　Abdullah1 N, Sulaiman1 F, Gerhauser H. Characterisation of oil palm empty fruit bunches for fuel application. Journal of Physical Science, 2011; 22: 1–24.

[6]　Amin NAS, Ya'aini N, Misson M, Haron R, Mohamed M. Enzyme pretreatment empty palm fruit bunch for biofuel production. Journal of Applied Science 2010; 10: 1181-1186 .

[7]　Syafwina, Honda Y, Watanabe T, Kuwahara M. Pre-treatment of oil palm empty fruit bunch by white-rot fungi for enzymatic saccharification. Wood Research 2002; 89: 19-20.

[8]　Millati R, Wikandari R, Trihandayani ET, Cahyanto MN, Taherzadeh MJ, Niklasson C. Ethanol from oil palm empty fruit bunch via dilute-acid hydrolysis and fermentation by Mucor indicus and Saccharomyces cerevisiae. Agriculture Journal 2011; 6: 54-59.

[9]　Han MH, Kim Y, Kim SW, Choi G-W. High efficiency bioethanol production from OPEFB using pilot pretreatment reactor. Journal of Chemical Technology and Biotechnology 2011; 86: 1527-1534.

[10]　Kim S, Park JM, Seo J-W, Kim CH. Sequential acid-/alkali-pretreatment of empty palm fruit bunch fibre. Bioresource Technolog 2012; 109: 229-233.

[11]　Misson M, Haron R, Kamaroddin MFA, Amin NAS. Pretreatment of empty palm fruit bunch for production of chemicals via catalytic pyrolysis. Bioresource Technolology 2009; 100: 2867 – 2873.

[12]　Jung YH, Kim IJ, Han J-I, Choi I-G, Kim KH. Aqueous ammonia pretreatment of oil palm empty fruit bunches for ethanol production. Bioresource Technology 2011; 102: 9806-9809.

[13]　Umikalsom MdS, Ariff AB, Karim MIA. Saccharification of pretreated oil palm empty fruit bunch fibre using cellulase of Chaetomium globosum. Journal of Agriculture and Food Chemistry 1998; 46: 3359-3364.

[14]　Lau MJ, Lau MW, Gunawan C, and Dale BE. Ammonia fibre expansion (AFEX) pretreatment, enzymatic hydrolysis, and fermentation on empty palm fruit Bunch fibre (EPFBF) for cellulosic ethanol production. Applied Biochemistry and Biotechnology 2010; 162: 1847-1857.

[15] Bahrin EK, Baharuddin AS, Ibrahim MF, Razak MNA, Sulaiman A, Abd-Aziz S, Hassan MA, Shirai Y, Nishida H. Physicochemical property changes and enzymatic hydrolysis enhancement of oil palm empty fruit bunches treated with superheated steam. BioResource 2012; 7: 1784-1801.

[16] Uemura Y, Omar WN, Tsutsui T, Bt Yusup S. Torrefaction of oil palm wastes. Fuel 2011; 90: 2585-2591.

[17] Uemura Y, Omar W, Othman NA, Yusup S, Tsutsui T. Torrefaction of oil palm EFB in the presence of oxygen. Fuel, 2011, doi 10.1016/j.fuel.2011.11.018.

[18] Sulaiman F, Abdullah N. Optimization conditions for maximising pyrolysis liquids of oil palm empty fruit bunches. Energy 2011; 36: 2352-2359.

[19] Genting Group Unveils Malaysia's First Commercially Produced Bio-oil using Breakthrough Technology. Report by Genting Group, Press Release; 21 August, 2005.

[20] Balat M, Kırtay E, Balat H. Main routes for the thermo-conversion of biomass into fuels and chemicals. Part 2. Gasification systems. Energy Conversion and Management 2009; 50: 3158–3168.

[21] Panwar NL, Kothari R, Tyagi VV. Thermo chemical conversion of biomass – Eco friendly energy routes. Renewable and Sustainable Energy Reviews 2012; 16: 1801-1816.

[22] Ogi T, Nakanishi M, Fukuda Y, Matsumoto K. Gasification of oil palm residues (empty fruit bunch) in an entrained-flow gasifier. Fuel, 2010. DOI: 10.1016/j.fuel.2010.08.028.

[23] Lahijani P, zainal ZA. Gasification of palm empty fruit bunch in a bubbling fluidized bed: A performance and agglomeration study. Bioresource Technology 2011; 102: 2068-2076.

[24] Mohammed MAA, Salmiaton A, Wan Azlina WAKG, Mohammad Amran MS, Fakhru'l-Razi A. Air gasification of empty fruit bunch for hydrogen-rich gas production in a fluidized-bed reactor. Energy Conversion and Management 2011; 52: 1555-1561.

[25] Mohammed MAA, Salmiaton A, Wan Azlina WAKG, Mohamad Amran MS. Gasification of oil palm empty fruit bunches: A characterization and kinetic study. Bioresource Technology 2011; 110: 628-636.

[26] Kassim MA, Kheang LS, Bakar NA, Aziz AA, Som RM. Bioethanol production from enzymatically saccharified empty fruit bunches hydrolysate using Saccharomyces cerevisiae. Research Journal of Environmental Sciences 2011; 5: 573-586.

[27] Noomtim P, Cheirsilp B. Production of butanol from palm empty fruit bunches hydrolysate by Clostridium acetobutylicum. Energy Procedia 2011; 9: 140-146.

[28] Ibrahim MF, Aziz SA, Razak MNA, Phang LY, Hassan MA. Oil palm empty fruit bunch as alternative substrate for acetone-butanol-ethanol production by Clos-tridium butyricum EB6. Applied Biochemistry and Biotechnology 2012; 166: 1615-1625.

[29] Nieves DC, Karimi K, Horváth IS. Improvement of biogas production from oil palm empty fruit bunches (OPEFB). Industrial Crops and Products 2011; 34: 1097-1101.

[30] O-Thong S, Boe K, Angelidaki. Thermo[philic anaerobic co-digestion of oil palm empty fruit bunches with palm oil mill effluent for efficient biogas production. Applied Energy 2012; 93: 648-654.h

Chapter 5

Hydrogen Conversion in DC and Impulse Plasma-Liquid Systems

Valeriy Chernyak, Oleg Nedybaliuk, Sergei Sidoruk,
Vitalij Yukhymenko, Eugen Martysh,
Olena Solomenko, Yulia Veremij, Dmitry Levko,
Alexandr Tsimbaliuk, Leonid Simonchik,
Andrej Kirilov, Oleg Fedorovich, Anatolij Liptuga,
Valentina Demchina and Semen Dragnev

Additional information is available at the end of the chapter

1. Introduction

It is well known [1] that hydrogen (H_2) as the environmentally friendly fuel is considered to be one of the future most promising energy sources. Recently, interest in hydrogen energy has increased significantly, mainly due to the energy consumption increase in the world, and recent advances in the fuel cell technology. According to the prognosis, in the next decades, global energy consumption will be increased by 59%, and still most of this energy will be extracted from the fossil fuels. Because of the traditional fossil fuels depletion, today there's a growing interest in renewable energy sources (f.e. – bioethanol, biodiesel). Bioethanol can be obtained from the renewable biomass, also it can be easily and safely transported due to its low toxicity, but it's not a very good fuel. Modern biodiesel production technologies are characterized by a high percentage of waste (bioglycerol) which is hard to recycle.

It is common knowledge [2] that addition of the syn-gas to the fuel (H_2 and CO) improves the combustion efficiency: less burning time, rapid propagation of the combustion wave, burning stabilization, more complete mixture combustion and reduction of dangerous emissions (NO_x). Besides, the synthesis gas is an important stuff raw for the various materials and synthetic fuels synthesizing. There are many methods of synthesis gas (including hydrogen) production, for example – steam reforming and partial liquid hydrocarbons oxidation. Also, there is an alternative approach – biomass reforming with low-temperature plasma assistance. Plasma is a very powerful source of active particles (electrons, ions, radicals, etc.), and therewith it can be catalyst for the various chemical processes activation. However, a major disadvantage of chemical processes plasma catalysis is weak processes control.

There is a bundle of electrical discharges that generate both equilibrium and non equilibrium plasma. For plasma conversion – arc, corona, spark, microwave, radio frequency, barrier and other discharges are used. One of the most effective discharges for the liquid hydrocarbons plasma treatment is the "tornado" type reverse vortical gas flow plasma-liquid system with a liquid electrode ("TORNADO-LE") [3]. The main advantages of plasma-liquid systems are – high chemical plasma activity and good plasma-chemical conversions selectivity. It may guarantee high performance and conversion efficiency at the relatively low power consumption. Moreover, those are systems of atmospheric pressure and above, and this increases their technological advantages.

Also, syn-gas ratio – hydrogen and carbon monoxide concentration ratio should be mentioned. As well, it should be taken into consideration that for efficient combustion (in terms of energy) of the synthesis gas it should contain more hydrogen, and in the case of the synthesis materials – they should contain more CO.

Relatively new possible solution to this problem – carbon dioxide recycling. Many modern energy projects have difficulties with the large amount of CO_2 storing and disposing. And it is also known that the addition of CO_2 to plasma during the hydrocarbons reforming may help to control plasma-chemical processes [4]. That is why the objective of the research is to study the influence of different amounts of CO_2 in the working gas on the plasma-chemical processes during the hydrocarbons conversion.

This research deals with hydrocarbons (bioethanol, bioglycerol) reforming by means of the combined system, which includes plasma processing and pyrolysis chamber. As a plasma source the "tornado" type reverse vortical gas flow plasma-liquid system with liquid electrode has been used [5].

Qualitatively new challenge is connected with a selectivity of the plasma chemistry strengthening by the transition of the chemical industry to "green chemistry". The last is a transition from the traditional concept of evaluating the effectiveness of the chemical yield to the concept that evaluates the cost-effectiveness as the exclusion of hazardous waste and non-toxic and/or hazardous substances [6].

A quantitative measure of the environmental acceptability of chemical technology is the ecology factor, which is defined as the ratio of the mass of waste (waste) to the mass of principal product. Waste is all that is not the principal product.

By the way, the most promising approaches in green chemistry is the implementation of processes in supercritical liquids (water, carbon dioxide) [7].

Water in supercritical condition unlimitedly mixes with oxygen, hydrogen and hydrocarbons, facilitating their interaction with each other - oxidation reactions are very fast in scH_2O (supercritical water). One particularly interesting application of this water - efficient destruction of chemical warfare agents. When mixed with other substances scH_2O can be used not only for oxidation but also in the reactions of hydrolysis, hydration, the formation and destruction of carbon-carbon bonds, hydrogenation, and others.

Besides, the use of pulsed electrical discharges in the liquid brings up new related factors: strong ultraviolet emission and acoustic or shock waves. In literature it can be found that systems with energies more than 1 kJ/pulse, that have negative influence on the lifetime of such systems. Reasonable from this perspective is the usage of pulsed systems with relatively low pulse energy and focusing of acoustic waves. In addition, the acoustic oscillations in such systems can be used as an additional mechanism of influence on chemical transformations.

In using of acoustic oscillations for chemical reactions the most attention is paid to systems with strong convergent waves. However, the processes during the collapse of the powerful convergent waves are studied unsufficiently. In the literature the systems of cylindrical, spherical or parabolic surfaces used in the focusing of shock waves for technological needs are known [8]. However, among their disadvantages should be noted that partial usage of the energy of acoustic wave and the problem of it's peripheral sources synchronization, which leads to distortion of the shock wave front ideality and reduces the focusing effectiveness.

Probably, more perspective method of using acoustic waves is their generation by single axial pulse electric discharge with further reflection from an ideal cylindrical surface. This approach can provide better symmetry of compression by convergent acoustic wave both in the gas and in the liquid. Probably, such mechanism can be exploited for scH_2O production

In addition, the re-ignition of electrical discharge at the moment of collapse convergent acoustic waves can lead to the plasma temperature increasing due to compression of the discharge channel, as well as the appropriate amplification of acoustic waves after the collapse.

It's clear that plasma-liquid systems (PLS) mentioned above have some sharp differences. Therefore, the first section of this article presents the results of our research on the addition of CO_2 to the "TORNADO-LE". And the second section of the article is devoted to investigation of double-impulse system in underwater electric discharge.

2. Bioethanol and bioglycerol conversion in "tornado" type plasma-liquid system with the addition of CO_2

2.1. Experimental set up

The experimental setting is shown in Fig. 1. Its base is a cylindrical quartz chamber (1) with diameter of 90 mm and height of 50 mm. Top (2) and bottom (3) it is hermetically closed with metal flanges. Camera is filled with fluid (4), the level of which has been maintained by the injection pump through the hole (5). Bottom flange is made of stainless steel. The stainless steel T-shaped cylindrical electrode (6), cooled with water, immerses in the liquid through the central hole in the bottom flange. There is a 5 mm thick metal washer on its surface (7) in the middle of which there is a hole in diameter of 10 mm. Sharp corners are rounded. This washer is used for reducing the waves (which have been moving to the quartz wall) amplitude on the liquid surface.

The top flange, made from duralumin, contains copper sleeve (13) with a diameter of 20 mm is placed in the center (2), and plays the role of the second electrode. The nozzle with diameter of 4 mm and length of 6 mm is located in the center of the copper sleeve (8). Gas is introduced into the flange (2) through the aperture (9). Gas flow changes the direction at 90 degrees inside the flange and injects tangentially into the channel (10). (10) The gas is rotated in the circular channel. Rotating gas (11) lands on the surface liquid and moves to the central axis of the system, where fells into the quartz cell through the nozzle (14), forming a plasma torch (12). Camera (14), in its turn, plays a role of pyrolytic chamber. Flow rate reaches the maximum value near the nozzle. Due to this, the zone of lower pressure is formed in the center of the gas layer, compared to the periphery. The conical structure appears over the liquid's surface near the system axis (Fig. 1). External static pressure is 1 atm. and internal - 1.2 atm (during discharge burning). Gas from quartz chamber (14) gets into the refrigerator (15), which is cooled with water at room temperature.

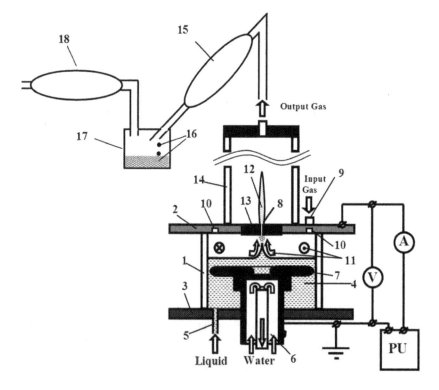

Figure 1. Schematic set up of the "TORNADO-LE".

Condensed matter (16) together with the gas from the refrigerator gets to the chamber (17). At the chamber exit (17) there's a flask (18), where gas is gathered for its composition diagnostics by means of mass spectrometry and gas chromatography. Study of plasma parameters is performed by emission spectrometry. The emission spectra registration procedure uses the system which consists of optical fiber, the spectral unit S-150-2-3648 USB, and the computer. Fiber is focusing on the sight line in the middle between the top flange (2) and the surface of the liquid (4).

The spectrometer works in the wavelength range from 200 to 1100 nm. The computer is used in both control measurements process and data processing, received from the spectrometer.

The voltage between the top flange and electrode, immersed in the liquid, is supplied by the power unit "PU". DC voltage provided is up to 7 kV. Two modes of operation have been considered:

1. "liquid" cathode (LC) – electrode immersed in the liquid has "minus" and the top flange has "plus";

2. solid" cathode (SC) - with the opposite polarity.

Electrode which has "plus" is grounded. Breakdown conditions are controlled by three parameters: the fluid level, the gas flow value and the voltage magnitude between the electrodes. The several modes of operation have been studied:

1. Various air flow and CO_2 ratio;

2. Discharge voltage varied within $U_d = 2.2 \div 2.4$ kV;

3. Discharge current varied within $I_d = 220 \div 340$ mA (ballast resistance hasn't been used).

At first, for the analysis of the plasma-chemical processes kinetics the distilled water (working fluid), and ethanol (ethyl alcohol solution in distilled water with a molar ratio $C_2H_5OH/H_2O = 1/9.5$), as a hydrocarbon model have been used. As the working gas mixture of air with CO_2, in a wide range of air flow and CO_2 ratios has been used. The ratio between air and CO_2 in the working gas changes in the ranges: $CO_2/Air = 1/20 \div 1/3$ for the working fluid C_2H_5OH/H_2O (1/9.5) and $CO_2/Air = 0/1 \div 1/0$ (by pure air to pure CO_2) - for distilled water.

Plasma component composition and population temperature of the excited electron (T_e^*), vibration (T_v^*) and rotational (T_r^*) levels of plasma components and relative concentrations of these components have been determined by the emission spectra. For the temperature population determination of the excited oxygen atoms electron levels the Boltzmann diagram method has been used [5]. T_e^* of oxygen atoms has been determined for the three most intense lines (777.2 nm, 844 nm, 926 nm). Temperature population of excited hydrogen atoms electron levels has been determined by the two lines of 656.3 nm and 486 nm relative intensities.

The effect of the presence of CO_2 in the system on the initial gas products has been investigated by means of "TORNADO-LE" current-voltage characteristics with changes in the working gas composition. T_v^* and T_r^* have been determined by comparing the experimentally measured emission spectra with the molecules spectra simulated in the SPECAIR program [9]. With help of this program and measured spectra, relative component concentrations in plasma have been determined. Also, the concentration of atomic components has been obtained by calculating the amount of oxygen that fell into a working system with the working gas flow. The hydrogen amount has been received from the electrolysis calculations. The output gas in reforming ethanol has been analyzed by gas chromatography and infrared absorption.

2.2. Results

The process of discharge ignition occurred as follows: the chamber is filled with liquid to a fixed level (5 mm above the washer). At the next stage a certain amount of gas flow forms the stationary cone from liquid; the voltage applied between the top flange and electrode immersed in a liquid starts gradually increase. When the voltage reaches a break-out value - U_b, a streamer appears for the first time. After that, burning discharge starts in split second, and then voltage decreases and current increases. After a second or two it is stabilized. During this time – static pressure rises inside the chamber from 1 to 1.2 atm. If you maintain the liquid level fixed, then the discharge is quite steady.

Liquid layer thickness of 5 mm has been chosen because that is the minimum liquid thickness in which the discharge burns between the liquid surface and the top flange. If the thickness is smaller plasma pushes the water toward the electrode immersed in the liquid and the discharge starts burning between two metal electrodes. Discharge goes into the arc regime. When the thickness of the distilled water layer above the washer is 5 mm (in the case of air flow only) break voltage reaches 4.5 kV and for a CO_2 flow - 6 kV. It is known [10], this increase in break-out voltage derives from the appearance of an additional loss channel of electrons – due to their sticking onto CO_2 molecules. This sticking has dissociative character and it is accompanied by the energy expense.

For example, the threshold reaction with CO_2 is 3.85 eV. Therefore CVC in pure CO_2 is decreased (Fig. 2). When the thickness of the C_2H_5OH/H_2O (1/9.5) solution layer above the washer is 5 mm (in the case of air flow only) the break voltage is 5.5 kV, and for the air flow mixture with CO_2 (CO_2/Air = 1/3) – 6.5 kV. Adding CO_2 to the air leads to the increase in the break-out voltage value. Adding ethanol to distilled water ($C_2H_5OH/H_2O = 1/9.5$) results in the increase of break voltage on 1 kV. Power supply unit provides maximum voltage of 7 kV. Increasing the thickness of the fluid layer above the washer (> 5 mm) leads to the increase of the break-out voltage value. There is no discharge ignition with a break-out voltage value of more than 7 kV. Therefore, 5 mm thickness of the liquid layer above the surface immersed in a liquid metal electrode (washer) has been chosen as the optimum one.

The current-voltage characteristics of the discharge are shown for the SC mode (Fig. 2 a; 2b). The cell has been filled with distilled water (Fig. 2a) or bioethanol (Fig. 2b).

The "tornado" type reverse vortex gas flow is formed by gas flow, which is a mixture of air with CO_2 in varying proportions. Ratio of CO_2/Air is changed in the range from 1/20 to 1/3, and in the case of ethanol and 1/0 in the case of water. Current varied in the range from 230 to 400 mA. The initial level of the working liquid is the same in all cases.

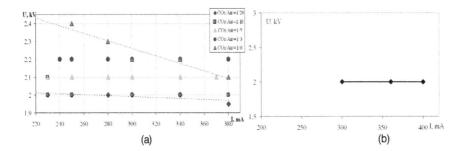

Figure 2. a) Current-voltage characteristics of the discharge at different ratios of CO_2/Air in the working gas. Working liquid - distilled water. Airflow - 55 and 82.5 cm³/s, the flow of CO_2 - 4.25, 8.5 and 17 cm³ /sec. b) Current-voltage characteristics of the discharge at the ratio CO_2/Air = 1/5 in the working gas. Working liquid – C_2H_5OH/H_2O (1/9.5) solution. Airflow - 82.5 cm³/s, the flow of CO_2 - 17 cm³/s.

The current-voltage characteristics show that adding a small amount of CO_2 (near 20%) to the working gas has no effect on the discharge type in various studied working liquids. In the range of flow ratios CO_2/Air from 1/20 to 1/5 characteristics are straight lines. It was observed that the increasing of CO_2 share in working gas causes discharge voltage supply rise.

Typical emission spectra of the plasma are shown in Fig. 3a and Fig. 3b for the cases with the distilled water as the working liquid and the solution of C_2H_5OH/H_2O (1/9.5).

The emission spectra show that when the working liquid is distilled water, plasma contains the following components: atoms H, O, and hydroxyl OH. In case when the working liquid is C_2H_5OH/H_2O solution (1/9.5), plasma has the following components: atoms – N, O, C, Fe, Cr, molecules – OH, CN, NH. The emission spectra shows that the replacement of the working liquid with distilled water with ethanol CN and lower electrode material made of stainless steel (anode) occur in plasma. Occasionally, during discharge burning breakdown may occur in C_2H_5OH/H_2O layer solution (1/9.5).

Those breakdowns may occur due to the fact that during the discharge burning, thickness of liquid layer, when the working fluid has a significant share of C_2H_5OH, a current channel is formed through the liquid layer to the metal electrode. And in the case of distilled water - plasma channel discharge ends near the surface of the liquid. It may indicate the presence of large liquid surface charge.

It was observed that the increase of CO_2 in the working gas (CO_2/air > 0.3) leads to an increase in the intensity of hydrogen and oxygen radiation lines (H and O) at the time when the intensity of the molecular component (OH) radiation, within the error, is stable (I = 300 mA, U = 1.9-2.4 kV, air flow 0 - 82.5 cm³/sec, the flow of CO_2 - 4.25 - 85 cm³/sec).

Fig. 4 shows the ratio of the hydrogen (Hα λ = 656.3 nm) and oxygen (O, λ = 777.2 nm) radiation intensity to the highest point of band hydroxyl (OH, λ = 282.2 nm) small intensity at different ratios of CO_2/air (I = 300 mA, U = 1.9 – 2.4 kV, air flow - 27.5, 55 and 82.5 cm³/s, the flow of CO_2 - 4.25, 8.5, 17, 42.5 and 85 cm³/s). High intensity bands haven't been used in the calculations because of the possible reabsorption. (I = 300 mA, U = 2 – 2.2 kV). In the case of distilled water

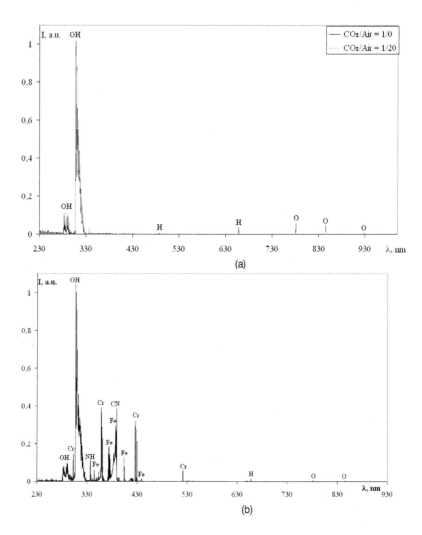

Figure 3. a) Emission spectrum of the plasma in TORNADO-LE plasma-liquid system, where the working liquid is distilled water. Working gas - a mixture of CO_2/air = 1/0, Id = 300 mA, U = 2.2 kV, the flow of CO_2 - 85 cm³/s and CO_2/air = 1/20, Id = 300 mA, U = 1.9 -2.0 kV, air flow - 82.5 cm³/s, the flow of CO_2 - 4.25 cm³/s. b) Emission spectrum of the plasma in the TORNADO-LE plasma-liquid system, where the working fluid is bioethanol. Working gas - a mixture of CO2/air = 1/20, Id = 300 mA, U = 2 kV, air flow - 82.5 cm3/sec, the flow of CO2 - 4.25 cm3/sec.

(Fig. 4a), results are presented for the three air flows - 27.5, 55 and 82.5 cm³/s and five CO_2 streams - 4.25, 8.5, 17, 42.5 and 85 cm³/s (I = 300 mA , U = 1.9 – 2.4 kV). Air and CO_2 flows are variated so that the total flow compiles similar values and achieves ratios of CO_2/air in a wide range from 1/20 to 1/0.

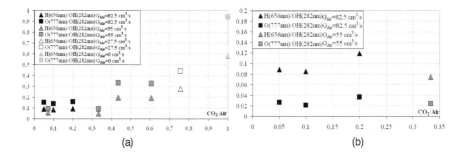

Figure 4. The ratio of the radiation intensity of hydrogen (H, λ = 656.3 nm) and oxygen (O, λ = 777.2 nm) to the peak of the band hydroxyl (OH λ = 282,2 nm) at different ratio CO_2/Air in the working gas. Working liquid - distilled water (a) I = 300 mA, U = 1.9 – 2.4 kV and bioethanol (b) I = 300 mA, U = 2 -2.2 kV.

In calculating the relative concentration ratio of hydrogen to oxygen from the emission spectra, it was observed that the hydrogen concentration is two times as much of the oxygen concentration for the case of distilled water as the working liquid - (I = 300 mA, U = 1.9 – 2.4 kV, airflow - 27.5, 55 and 82.5 cm^3/sec, the flow of CO_2 - 4.25, 8.5, 17, 42.5 and 85 cm^3/sec), and ten times as much when the working liquid is C_2H_5OH/H_2O solution (1/9.5) (I = 300 mA, U = 2 - 2.2 kV, air flow - 55 and 82.5 cm^3/sec, the flow of CO_2 - 4.25, 8.5 and 17 cm^3/sec). However, according to the calculations, these components production by means of electrolysis and their extraction from the working gas, the oxygen concentration exceeds the average hydrogen concentration in three orders of magnitude, unless the case when the pure CO_2 is used as a working gas.

It should be noted that the addition of CO_2 reduces the discharge stability, especially in the case of bioethanol. In determination of the temperature population excited electron levels of plasma atomic component the most intense lines (spectra with the smallest possible accumulation in the experiment measurement of 500 ms) have been used, according to the discharge burning particularity. Also, it affects the parameters determination accuracy.

Temperature of excited hydrogen electron population levels is determined by the relative intensities (two lines of 656 nm and 486 nm). For the case where the working liquid is distilled water – T_e^*(H) = 5500 ± 700 K (I = 300 mA, U = 1.9-2.4 kV, air flow - 27.5, 55 and 82.5 cm^3/s, the flow of CO_2 - 4.25, 8.5, 17, 42.5 and 85 cm^3/s) and as for the bioethanol T_e^*(H) = 6000 ± 500 K - (I = 300 mA, U = 2 – 2.2 kV, air flow - 55 and 82.5 cm^3/s, the flow of CO_2 - 4.25, 8.5 and 17 cm^3/s). Also, the oxygen T_e^*(O) has been defined by the Boltzmann diagrams method, in the case of distilled water. The three most intense lines (777.2 nm, 844 nm, 926 nm) are used in this method. So, we have T_e^*(O) = 4700 ± 700 K.

Temperatures of OH excited vibrate and rotational population levels have been determined by comparing the experimentally measured emission spectra with the molecular spectra modeled in The SPECAIR program. In the case when the working liquid is distilled water, appropriate temperatures are: T_r^*(OH) = 3000 ± 1000 K, T_v^*(OH) = 4000 ± 1000 K (I = 300 mA,

U = 1.9-2.4 kV, air flow - 27.5, 55 and 82.5 cm3 / s, the flow of CO_2 - 4.25, 8.5, 17, 42.5 and 85 cm^3/s). Also, population temperatures of vibration and rotational levels for OH and CN have been determined in case of C_2N_5ON/H_2O (1/9.5) solution as the working liquid, they are: T_r^* (OH) = 3500 ± 500 K, T_v^*(OH) = 4000 ± 500 K, T_r^*(CN) = 4000 ± 500 K, T_v^*(CN) = 4500 ± 500 K) (I = 300 mA, U = 2 - 2.2 kV, air currents - 55 and 82.5 cm^3/s, the flow of CO_2 - 4.25, 8.5 and 17 cm^3/s). Temperatures for other molecular components haven't been determined because of their bands low intensity.

During the study, it turned out that the addition of CO_2 weakly affects the population temperature of excited electron, vibration and rotational levels of plasma components (Fig. 5) (I = 300 mA, U = 1.9 - 2.4 kV, air flow - 27.5, 55 and 82.5 cm^3/s, the flow of CO_2 - 4.25, 8.5, 17, 42.5 and 85 cm^3/s). Weak tendency to temperature decrease has been observed, but these changes do not exceed the error.

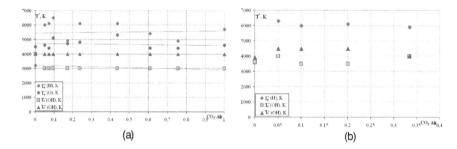

Figure 5. Population temperatures of excited electron, vibration and rotational levels of plasma components at different ratio of CO_2/Air in the working gas. Working liquids - distilled water (a) and ethanol (b)

Fig. 6–7. shows the results of gas chromatography bioethanol conversion output products. Results are presented for two air streams 55 and 82.5 cm^3/s + three CO_2 streams: 4.25, 8.5 and 17 cm^3/s (I = 300 mA, U = 2 – 2.2 kV). CO_2/Air ratio in the range from 1/20 to 1/3 has been changing exactly this way. Selection of gas into the flask has been taken place at the refrigerator output. The flask has been previously pumped by the water-jet pump to the pressure of saturated water vapor (23 mm Hg).

Fig. 6 shows the gas chromatography comparison of bioethanol conversion output products with and without the addition of CO_2. The air flow is constant – 55 cm^3/s, in case of CO_2/Air = 1/3 – 17 cm^3/s of CO_2 has been added to the air (the total flow has been increased, which may explain the decrease in the percentage of nitrogen at a constant air flow; I = 300 mA, U = 2 - 2.2 kV). This histogram shows that adding of carbon dioxide leads to a significant increase of the H_2 component percentage, CO (syn-gas) and CH_4 in the output gas. This may indicate that the addition of CO_2 during the ethanol reforming increases the conversion efficiency, because CO_2 plays a burning retarder role.

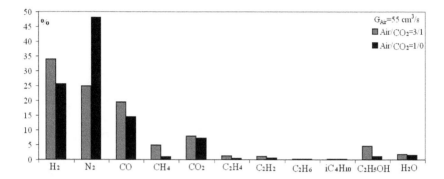

Figure 6. Gas chromatography comparison of bioethanol conversion output products with and without the addition of CO_2.

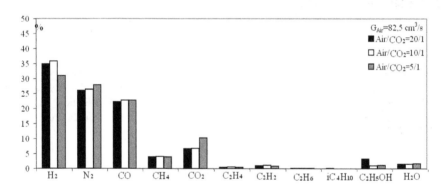

Figure 7. Gas chromatography comparison of bioethanol conversion output products by adding different amounts of CO_2.

The ethanol solution consumption for the SC mode with current of 300 mA and air flow of 55 cm^3/ equals 6 ml/min, and for the air flow of 82.5 cm^3/s and CO_2 of 17 cm^3/s mixture - 10 ml/min.

According to the gas chromatography, in the studied correlations range of CO_2/Air, syn-gas ratio ($[H_2]/[CO]$), changes slightly – look at Fig. 8. Measurements were made by two air streams of 55 and 82.5 cm^3/s and three CO_2 streams of – 4.25, 8.5 and 17 cm^3/s; I = 300 mA, U = 2 – 2.2 kV.

Besides the gas chromatography, the output gas composition has been studied by means of infrared spectrophotometry (IRS). Fig. 9 shows a typical IRS spectrum of the output gas. In the SC mode (current 300 mA, voltage 2 kV) the working liquid is ethyl alcohol and distilled water mixture (C_2H_5OH/H_2O = 1/9.5), and the working gas – air (82.5 cm^3/s) and CO_2 (4.25 cm^3/s) mixture. Research has been carried out in a ditch with a length of 10 cm and a diameter of 4 cm. Pressure inside the ditch has been 1 atm. The ditch walls have been made of BaF_2.

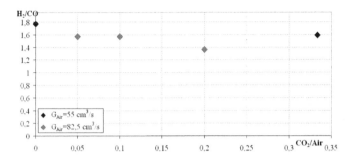

Figure 8. Syn-gas ratio of bioethanol conversion output products for various ratios of CO_2/Air in the range between 0/1 - 1/3.

Fig. 10 shows the dependence of the CO transmission standardized maximum intensity peaks (2000 - 2250 cm^{-1}) in the syn-gas, depending on the CO concentration according to gas chromatography results. Standardization has been conducted for the maximum intensity value of the CO transmission peak bandwidth at the SC mode with the current of 300 mA, voltage - 2 kV, the mixture of ethyl alcohol and distilled water (C_2H_5OH/H_2O = 1/9.5) as the working liquid, and the mixture of air (82.5 cm^3/s) and CO_2 (4.25 cm^3/s), as the working gas.

According to IR spectrophotometry CO fraction in the synthesis gas is practically the same. According to gas chromatography CO fraction in the synthesis gas in the different operation modes stays on the same level as well (the changes are in the range of 1%). So IR spectrophotometry can be used determine the composition of synthesis gas under ethanol reforming.

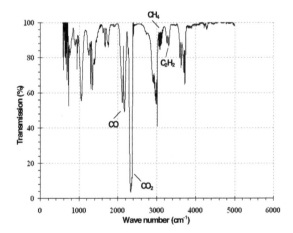

Figure 9. The SC mode (current 300 mA, voltage 2 kV) the working liquid is ethyl alcohol and distilled water mixture (C_2H_5OH/H_2O = 1/9.5), the working gas – air (82.5 cm^3/s) and CO_2 (4.25 cm^3/s) mixture

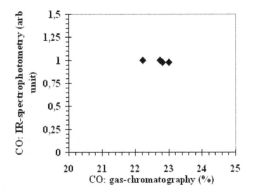

Figure 10. Dependence of the normalized maximum intensity peaks (2000-2250 cm⁻¹) transmission of CO in the syn-gas, depending on the concentration of CO according to gas chromatography data.

Plasma provides gas generation, which contains a certain amount of the syn-gas. The energy needed for this plasma support (Q_p) has been calculated by the following formula:

$$Q_p = P_d t, \tag{1}$$

where P_d - power that has been embedded into the discharge, t – production time of gas volume unit during the reforming process. Electrical energy transformation coefficient α has been calculated by the formula:

$$\alpha = \frac{Q_s}{Q_p} \tag{2}$$

where Q_s - energy that is released during the complete combustion of syngas (obtained in the reforming process).

Electrical energy transformation coefficient α has value of 0.81 for the "TORNADO-LE" with an ethanol solution and pure air flow 55 cm³/s. And the CO_2 addition (the ratio of CO_2/air = 1/3) gives the value of $\alpha = 1,01$. System electrical parameters are as follows: I = 300 mA, U = 2 – 2.2 kV.

2.3. Model and calculations

In the model of calculations was assumed that the discharge is homogeneous over the entire volume. It is justified at zero approximation, because the time of gas mixing in the radial direction is less than the times of characteristic chemical reactions. Also we neglect the processes in the transitive zone between the discharge to post-discharge. Thus, the time of gas pumping through the transition region is too short for the chemical reactions to have a sufficient influence on the concentration of neutral components.

The total time of calculation is divided into two time intervals: the first one is the calculation of the kinetic processes of fast generation of active atoms and radicals in the discharge region. Those components accelerate the formation of molecular hydrogen, carbon oxides and production of other hydrocarbons. The second time interval is the oxidation of the gas mixture in the post-discharge region as a result of the high gas temperature and the presence of O and OH. These components remain in the mixture after the dissociation of water and oxygen molecules by electron impacts in the plasma. The oxidation of generated hydrocarbons has a noticeable influence on kinetics in the investigated mixture due to the high gas temperature.

Under the aforementioned conditions, the characteristic time of oxidation is approximately equal to the air pumping time through the discharge region ($\sim 10^{-3}$–10^{-2} s). The following system of kinetic equations is used in order to account for the constant air pumping through the system:

$$\frac{dN_i}{dt} = S_{ei} + \sum_j k_{ij}N_j + \sum_{j,l} k_{ijl}N_jN_l + \cdots + K_i - \frac{G}{V}N_i - kN_i \tag{3}$$

N_i, N_j, N_l in the equation (3) are the concentrations of molecules and radicals; k_{ij}, k_{iml} are the rate constants of the processes for the i-th component. The rates of electron–molecule reactions S_{ei} are connected with discharge power and discharge volume. The last three terms in equation (1) describe the constant inflow and outflow of gas from the discharge region. The term K_i is the inflow of molecules of the primary components (nitrogen, oxygen, carbon dioxide, water and ethanol) into the plasma, G/VN_i and kN_i are the gas outflow as the result of air pumping and the pressure difference between the discharge region and the atmosphere. In order to define the initial conditions, the ethanol/water solution is assumed to be an ideal solution. Therefore, the vapor concentrations are linear functions of the ethanol-to-water ratio in the liquid. The evaporation rates K_i of C_2H_5OH and H_2O are calculated from the measured liquids' consumption. The inflow rates K_i of nitrogen and oxygen are calculated by the rate of air pumping through the discharge region:

$$K_i = \frac{G}{V}N_i^0 \tag{4}$$

where N_i^0 correspond to $[N_2]$ and $[O_2]$ in the atmospheric pressure air flow.

The gas temperature in the discharge region is taken to be constant in the model. In reality, the gas temperature T is dependent on the gas pumping rate and the heat exchange with the environment. Therefore, in order to take into account those influences, T is varied in the interval 800–2500K (similarly to the experimentally obtained temperature spread). After $\sim 10^{-2}$ s, the balance between the generation and decomposition of the components leads to saturation of concentrations of all species. This allows us to stop the calculations in the discharge region and to investigate the kinetics in the post-discharge region. System (3) is solved without accounting for the last three terms on the time interval without the plasma. The calculations are terminated when the molecular oxygen concentration reaches zero level.

The full mechanism developed for this experimental work is composed of 30 components and 130 chemical reactions between them and its closed to [11]. The charged particles (electrons and ions) are ignored in the mechanism, because of low degree of ionization of the gas (~ 10^{-6} – 10^{-5}). Nitrogen acts as the third body in the recombination and thermal dissociation reactions. In the non-equilibrium plasma almost the entire energy is deposited into the electron component. The active species, generated in the electron–molecular processes, lead to chain reactions with ethanol molecules.

Numerical simulation of kinetics showed that the main channels of H_2 generation in the plasma were ethanol abstraction for the first 10–100μs, and hydrocarbon abstraction afterwards. Additionally, the conditions when the reaction between H_2O and hydrogen atoms was the main channel of H_2 production were found. A kinetic mechanism, which adequately described the chemistry of the main components, was proposed. The model did not account for nitrogen-containing species, and nitrogen was considered only as a third body in recombination and dissociation reactions. The comparison between experiments and calculations showed that the mechanism can adequately describe the concentrations of the main components (H_2, CO, CO_2, CH_4, C_2H_4, C_2H_6, and C_2H_2).

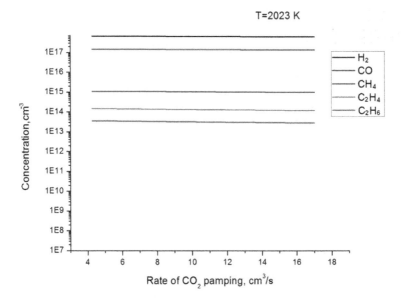

Figure 11. The dependence of the reaction main products of the flow rate of CO_2 (inside discharge), T = 2023 K

However, it should be noted that with the increase in temperature to 2523 K leads to the fact that the output of the reactor is not observed almost no light hydrocarbons. They simply "fall apart" and burned. That leaves the most stable elements such as H_2O, N_2, CO_2. This suggests that the increase in temperature up to these values is not advisable because of the decrease in the yield of useful products (see Fig. 11 and Fig. 12a,b).

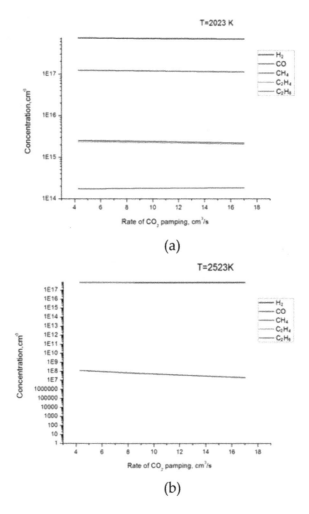

Figure 12. a). The dependence of the reaction main products of the flow rate of CO_2 (after discharge), T = 2023 K. b). The dependence of the reaction main products of the flow rate of CO_2 (after discharge), T = 2023 K

These calculations are based in good correspondence with the experimental data (see Fig. 8).

Other model hydrocarbon is bioglycerol (crude glycerol) which is a byproduct of the biodiesel manufacture. Biodiesel is a popular alternative fuel. It is carbon neutral, has emissions equivalent or below diesel, is biodegradable, non-toxic, and is significantly cheaper to manufacture than its petroleum equivalent. However there is one significant drawback: for every 10 gallons of biodiesel produced, roughly 1 gallon of bioglycerol is created as a byproduct.

Biodiesel is produced by mixing vegetable oil and potassium hydroxide KOH. Therefore, the large-scale production of environmentally friendly and renewable fuel may lead to possible bioglycerol accumulation in large quantities, which, in turn, can cause environmental problems, as it is comparably bad fuel. In addition, it has a rather large viscosity of 1.49 Pa•s, which is larger for almost three orders of magnitude than ethanol and water viscosity. The solution to this problem would be "TORNADO-LE" usage for bioglycerol reforming. Pure glycerol chemical formula is $C_3H_5(OH)_3$. However, bioglycerol contains various impurities (including a set of alkali).

Fig. 13 shows a photograph of burning discharge, where the working liquid is bioglycerol and working gas - air. Research is conducted by the SC polarity, because this mode has lowest liquid consumption.

Figure 13. Photo of the combustion discharge in which the working liquid is bioglycerol and working gas - air.

Fig. 14 shows the typical emission spectrum of the plasma discharge in a "TORNADO-LE" where the working liquid is bioglycerol doped with alkali. It is registered at a current of 300 mA, voltage – 2 kV, air flow – 110 cm³/s. Optical fiber is oriented on the sight line, parallel to the liquid surface in the middle of the discharge gap. The distance from the liquid surface to the top flange equals 10 mm.

Emission spectrum (Fig. 14) is normalized to the maximum Na doublet (588.99 nm, 589.59 nm). It contains K (404.41 nm, 404.72 nm, 766.49 nm, 769.89 nm), Na (588.99 nm, 589.59 nm), Ca (422.6 nm) lines, and a part of continuous spectrum, which indicates that the there's a soot in the discharge. Temperature, which is defined by the plasma continuous emission spectrum is 2700 ± 100 K.

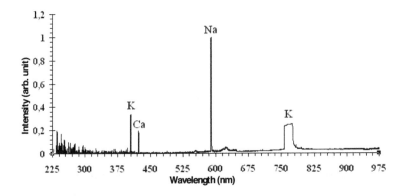

Figure 14. Typical emission spectrum of the plasma discharge, which burns in a mixture of air and bioglycerol / alkali.

The K, Na, Ca elements presence in the discharge gap complicates the plasma kinetics numeric modeling of the bioglycerol reform process. The gas flow rate at the system outlet is 190 cm³/s, i.e. by 80 cm³/s larger than the initial (110 cm³/s), which indicates bioglycerol reforming to the syn-gas. Liquid flow is 5 ml/min. Change of the CO_2 share in the working gas weakly affects the spectrum appearance.

Based on the continuous nature of the plasma emission spectra, we compared the experimental results with the calculated spectra of the blackbody radiation. Calculations have been performed by using Planck's formula.

Fig. 15 shows the computational grid with step of 200-300 K in the temperature range from 2500 K to 3500 K and the plasma emission spectrum in the case of bioglycerol, as a working fluid (air flow - 82.5 cm³/s, the flow of CO_2 - 17 cm³/s, CO_2/Air = 1/5, I_d = 300 mA, U = 600 V). All spectra are normalized to the intensity, which is located at a wavelength of 710 nm.

The data in Fig. 15 show that the plasma emission spectrum coincides with the calculated by the Planck formula for the temperature T = 2800 ± 200 K. Since bioglycerol contains alkali metals, which represent an aggressive environment, the gas chromatography can't be used. Therefore, in order to determine the gas composition, formed the bioglycerol reformation IR and mass spectrometry have been used.

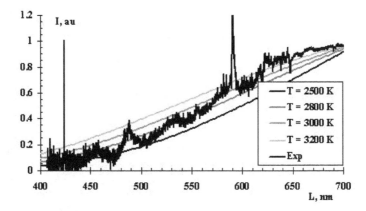

Figure 15. Plasma emission spectrum in the case when the working gas is a mixture $SO_2/Air = 1/5$ (air flow - 82.5 cm³/s, the CO_2 flow - 17 cm³/s), $I_d = 300$ mA, $U = 600$ V and calculated spectra of blackbody radiation)

With infrared transmission spectra one can see that the transition to bioglycerol increases the amount of such components as CO_2 (2250-2400 cm⁻¹), CO (2000-2250 cm⁻¹), CH_4 (3025-3200 cm⁻¹), C_2H_2 (3200-3350 cm⁻¹).

2.4. Discussions

Electrical energy is added to the "TONADO-LE" plasma-liquid system in the form of plasma power. Plasma acts as a catalyst and thus this power should be controlled. In addition to electric energy for plasma we incorporate hydrocarbon (ethanol or bioglycerol) as an input to the system. These hydrocarbons are raw material for syn-gas generation but they are also a fuel which has some energy associated with it. So, we input some energy to the system (hydrocarbon + electricity) and we get syn-gas, which is potentially a source of energy as well.

Carbon dioxide adding leads to a significant increase the percentage of $H_2 + CO$ (syn-gas) and CH_4 components in the exhaust. This may indicate that the CO_2 addition under the ethanol reforming increases the conversion efficiency, because CO_2 plays a role of the retarder in the system by reducing the intensity of the conversion components combustion.

The transmission spectra of infrared radiation indicate that the exhaust gas obtained by ethanol solution conversion, contains such components as CO, CO_2, CH_4, C_2H_2. It was found that CO_2 adding reduces the CH_4 and C_2H_2 amount, but does not affect the amount of producted CO.

The possibility of hydrocarbons reforming, which have considerable viscosity (bioglycerol) in the "TORNADO-LE" is shown. This gives a possibility to avoid environmental problems due to the bioglycerol accumulation during biodiesel production.

The α coefficient [see (2)] in bioglycerol reforming is higher than ethanol reforming at the same ratios of CO_2/Air in the input gas. This may be connected with the lower power consumption on the plasma generation in case of bioglycerol reforming. Bioglycerol contains alkaline dash, which increases the bioglycerol conductivity. Bioglycerol reforming products contain mainly CO and hydrocarbons CH_4, C_2H_2, which also gives some contribution to energy yield.

3. Dynamic impulse plasma–liquid systems

3.1. Experimental set up

The experimental setting is shown in Fig. 16. The main part of the system is cylinder with height H = 10 mm, and radius R = 135 mm. Its lateral surface made of stainless steel with a thickness of 5 cm. This cylinder is filled with liquid for experimental operations. The electrodes are placed perpendicular to the cylinder axis. They have the diameter of 10 mm, made of brass, and their ends are shaped hemispheres with a radius of curvature of 5 mm. The discharge (2) is ignited between the rounded ends of the electrodes. At a distance of 40 mm from the lateral surface of the cylinder is piezo-ceramic pressure sensor (3), which records acoustic vibrations in the fluid, caused by electric discharge under water. The distance between the sensor head and the system axis = L.

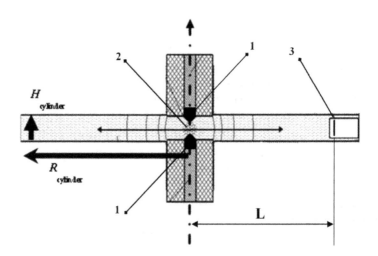

Figure 16. Schematic diagram of plasma-liquid system with a pulsed discharge, 1 - electrodes with brass tips, 2 – plasma, 3 – piezo-ceramic pressure sensor.

The cylindrical system could be located in a horizontal position (Fig. 17a) or vertical one (Fig. 17b). The full volume (0.5 l) of system is fluid-filled. The fluid in the system can be processed as in static mode (no flow), and dynamic one (with flow ~ 15 cm³/s). Additional supply of gas may be realized in the system also (airflow ~ 4 cm³/s), which is injected through a spray nozzle (source diameter 8 mm) located near the inner wall of the cylinder at a distance of 130 mm from the discharge gap (Fig. 16). The working fluids are: the tap water (with and without flow), distillate and ethanol (96%, no flow).

The main feature of electrical scheme for pulsed power feeding of discharge in a liquid is usage of two independent capacitors which are supplied two independent sources of power (1 kW). Pulsed discharge realized in two modes: single and double pulses. In the single pulse mode only one capacitor is discharged with a frequency of 0 - 100 Hz.

Double pulse mode is realized as follows: one capacitor discharges in the interelectrode gap through air spark gap; the clock signal from the Rogowski belt after first breakdown is applied to the thyratron circuit and second capacitor discharges through it. This set of events leads to the second breakdown of the discharge gap and second discharge appearance.

Delay of the second discharge ignition may be changed in range of 50 - 300 microseconds. The following parameters are measured: discharge current and the signal from the pressure sensor. The Rogowski belt has the sensitivity 125 A/V, and its signal is recorded with an oscilloscope. Capacity for the first discharge (C_1) = 0.105 µF and it is charged to U_1 = 15 kV (energy E_1 = 12 J), capacity for the second discharge C_2 = 0.105 µF and it is charged to U_2 = 18 kV (energy E_2 = 17 J).

A distance between electrodes can be changed in the range of 0.25 - 1 mm. The second discharge can be ignited at the moment (according to the delay tuning) when the reflected acoustic wave, created by the first electric discharge in liquid, returns to the center of the system (the time of its collapse ~ 180 ms).

(a)   (b)

Figure 17. Photograph of the cylinder from the outside: a) horizontal position, b) vertical position.

The composition of ethanol and bioethanol reforming products is studied with gas chromatography, in case of bioglycerol reforming - mass spectrometry and infrared spectrophotometry.

3.2. Results

Oscillograms of current and acoustic signal for different distances between electrodes (0.5 and 1 mm) are presented in Fig. 18. These oscillograms show the presence of electrolysis phase before breakdown, while duration of electrolysis increases with interelectrode distance.

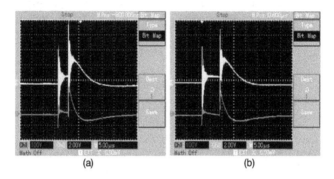

Figure 18. Oscillograms of the discharge current (top curve) and signal piezo-ceramic pressure sensor (lower curve): a) d = 0.5 mm, b) d = 1 mm. Tap water flow = 15 cm³/sec, without the input gas stream, C = 0.18 uF, U = 13.5 kV; ballast resistor in the discharge circle: R_b = 20 Ohm, the cylinder is in horizontal position.

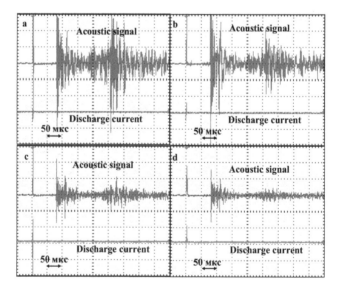

Figure 19. Oscillograms of current and acoustic signal in the single pulse mode at the different discharge ballast resistor: R_b: a) - 0 Ohm, b) - 10 Ohm; c) - 20 Ohm, d) - 50 Ohm. Tap water flow 15 cm³/s, without the input gas stream, d = 0.5 mm, C = 0.015 µF, U = 19.5 kV, the cylinder is in horizontal position.

Fig. 19 shows the acoustic signal dependence from ballast resistor in the discharge electric circuit. The acoustic signal has two splashes: №1 - the first diverging acoustic wave, and №2 - the second diverging acoustic wave. When the ballast resistor is increased, first and second acoustic signal splashes are decreased. This may be due to the fact: we increase the ballast resistor and set measures to the discharge current, as a result the injected into the discharge gap energy is diminished.

Also, there is a signal immediately behind the front of the first splash, which is founded in all cases at 110 microseconds interim from the beginning of the discharge current. The acoustic wave passes the way near 17 cm during this time. The pressure sensor is located at the distance of 2 cm from the lateral surface, so the acoustic signal passes the way near 12 cm to the sensor. Thus, there is a second stable signal after the first splash through time ~ 29 µs, which corresponds to the path ~ 4.4 cm, so the signal can be the convergent acoustic waves reflected from the wall.

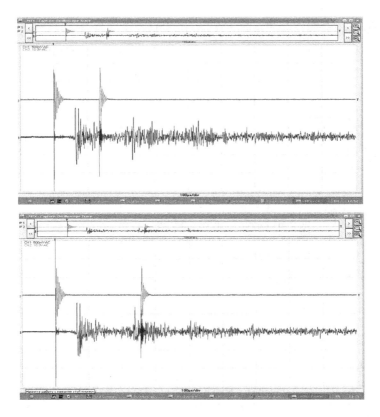

Figure 20. Oscillograms of the discharge current (top oscillogram) and acoustic signal (lower oscillogram) at different delays of the second discharge pulse.

There is the third acoustic signal splash in the experiment, but it does not affect the second discharge pulse delay in relation to the first. In addition, there is no acoustic signal from to the second discharge pulse in the double pulse mode, although the single pulse signal is present in the single pulse mode (Fig. 20).

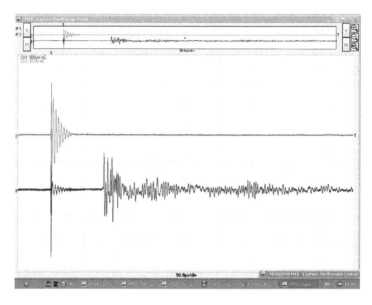

Figure 21. Oscillograms of the discharge current (top oscillogram) and the acoustic signal (lower oscillogram) in the single pulses mode. Working fluid - ethanol, d = 0.25 mm, $C_1 = 0.105$ μF, $U_1 = 15$ kV, the cylinder in the vertical position

Fig. 21 shows clearly that the duration and amplitude parameters for the first current pulse in the ethanol are virtually indistinguishable from the first current pulse in distilled water at any cylinder orientations. The ratio of the second acoustic signal amplitude to the first acoustic signal amplitude in the ethanol is noticeably less than in the tap water and distillate.

The results of oscillographic studies of the discharge current and acoustic signals in double pulses mode demonstrate that the first discharge in double pulses mode takes place in the narrow gas channel with a radius comparable to the size of the plasma channel, and the second discharge takes place in the wide channel with radius larger than the plasma channel.

Next, we present the results of ethanol reforming studies in the impulse plasma-liquid system with double pulses mode and their comparison with the results obtained for "TORNADO-LE".

The mass spectrometer studies of ethanol reforming in the impulse PLS of cylindrical geometry were carried out in the following modes: single pulse mode (C = 0.105 μF, U = 15 kV, f = 15 Hz, power 180 W) and double-pulse mode ($C_1 = C_2 = 0.105$ μF, $U_1 = 15$ kV, $U_2 = 15$ kV, f = 15 Hz, second pulse delay = 170 μs, this time is less on 10 μsec than collapse time, the power is 435

Wt), the interelectrode distance - 0.25 mm, working liquid - ethanol (96%), the input airflow is 4 cm³/s.

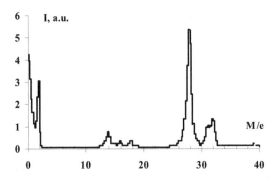

Figure 22. Mass spectrum for double pulse mode. Ethanol is without flow, inlet gas stream - 4 cm³/s, d = 0.25 mm, C_1 = C_2 = 0.105 μF, U_1 = 15 kV, U_2 = 18 kV, the cylinder is in the vertical position, f = 15 Hz.

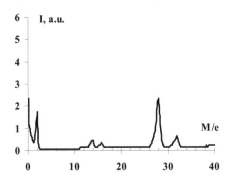

Figure 23. Mass spectrum for the single pulse mode. Ethanol without flow, inlet air flow - 4 cm³/s, d = 0.25 mm, C_1 = 0.105 μF, U1 = 15 kV, the cylinder is in the vertical position, f = 15 Hz.

The mass spectrometric studies show that the main components of the output fuel mixture are: hydrogen, carbon dioxide, and molecular nitrogen. The values of these components in the mixture: H_2 - 29%, CO - 17% for double pulse mode and H_2 - 35%, CO - 7% for single pulse mode. That is, with the same molecular hydrogen output, the carbon dioxide yield is significantly increased in double pulses mode.

The typical mass spectrum (Fig. 24) of the ethanol reforming (ethanol aqueous solution ethanol with concentrations 3.5, 13 and 26 percents) in the "TORNADO-LE". The power is 640 Wt. It is injected in the plasma for its generation, and inlet air flow is 55 cm³/s.

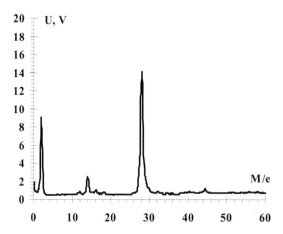

Figure 24. Mass spectrum of the output mixture in the ethanol reforming (ethanol - 26%) in "TORNADO-LE" PLS

The following Tab.1 shows the values ratio generating the volume unit of (H_2 + CO) mixture per unit of electrical power, which is injected into the plasma under reforming process in the impulse PLS of cylindrical geometry with double pulses mode, and in the "TORNADO-LE":

	Impulse PLS of cylindrical geometry with double pulses mode	"TORNADO-LE"
Single pulse	0.027 cm³/Wt	
Double pulses	0.0082 cm³/Wt	
Bioethanol 6,5%		0.0024 cm³/Wt
Bioethanol 13%		0.0079 cm³/Wt
Bioethanol 26%		0.0615 cm³/Wt

Table 1. The volume unit of (H_2 + CO) mixture per unit of electrical power in various PLS

The H_2 and CO components yield increases with increasing of the ethanol aqueous solution concentration. This concentration has maximum value 26%, and H_2 - 26%, CO - 14%. The results of these systems studies show, that the pressure, in region collapse of converging shock waves (with pulse energy > 10 J), exceeds critical (Tab. 2). So, the additional increase chemical activity due to supercritical processes inclusion can be achieved in this situation.

Solvent	Molecular mass	Critical temperature, T_{crit}	Critical pressure, P_{crit}	Critical density, ρ_{crit}
	g/mol	K	MPa (bar)	g/sm³
CO_2	44.01	303.9	7.38 (72.8)	0.468
H_2O	18.015	647.096	22.064 (217.755)	0.322
ethanol	46.07	513.9	6.14 (60.6)	0.276

Table 2. Critical parameters of different solvents

4. Discussion

The presence of electrolysis phase preceding electrical breakdown of heterophase environment demonstrates that the discharge development in the liquid perform with microbubbles. This result confirms the theory of "bubble" breakdown proposed by Mark Kushner [12].

The formation of convergent acoustic wave after reflection from the ideal solid cylindrical surface was investigated. It is shown that acoustic waves may be effectively focused during these waves passage inside the system.

The research of ethanol reforming in pulse plasma-liquid system has shown that transition from single pulse mode to double pulse mode is accompanied by reduction syn-gas ratio ($[H_2]/[CO]$).

When the working fluid is bioglycerol the K, Na, Ca lines are presented in emission spectra and there is a solid continuous spectrum, which indicates that microparticles are present in the discharge. Its temperature is $T = 2800 \pm 200$ K.

5. General conclusions

On the base of our results in bioethanol and bioglycerol CO_2-reforming by "TORNADO-LE" plasma-liquid system, we can say that:

1. This process has special features, connected with CO_2 retarding role in the conversion components combustion;

2. In this system there is the possibility of reforming of hydrocarbons with significant viscosity (such as bioglycerol);

3. All the diagnostic methods, used in the "TORNADO-LE" plasma-liquid system, indicate that there're no NO_x compounds in the bioethanol and bioglycerol reforming products.

The investigations of bioethanol and bioglycerol in pulse plasma-liquid system have shown:

1. The main components of the output fuel mixture in this case are: hydrogen, carbon dioxide, and molecular nitrogen, but the carbon dioxide yield is significantly increased in double pulses mode;

2. The formation of supercritical water in such system and its possible applications for recycling of organic waste and for nanocrystalline particles (in particular, oxide catalysts and other nanocrystalline materials, such as nanotubes) productions needs for additional researches.

Author details

Valeriy Chernyak[1*], Oleg Nedybaliuk[1], Sergei Sidoruk[1], Vitalij Yukhymenko[1], Eugen Martysh[1], Olena Solomenko[1], Yulia Veremij[1], Dmitry Levko[2,3], Alexandr Tsimbaliuk[2], Leonid Simonchik[4], Andrej Kirilov[4], Oleg Fedorovich[5], Anatolij Liptuga[6], Valentina Demchina[7] and Semen Dragnev[8]

*Address all correspondence to: chernyak_v@ukr.net

1 Taras Shevchenko National University of Kyiv, Ukraine

2 Institute of Physics, National Academy of Sciences of Ukraine, Kyiv, Ukraine

3 Physics Department, Technion, 32000, Haifa, Israel

4 B.I. Stepanov Institute of Physics, National Academy of Sciences, Minsk, Belorus

5 Institute of Nuclear Research, National Academy of Sciences of Ukraine, Kyiv, Ukraine

6 V.E.Lashkaryov Institute of Semiconductor Physics, National Academy of Science of Ukraine, Kyiv, Ukraine

7 The Gas Institute, National Academy of Science of Ukraine, Kyiv, Ukraine

8 National University of Life and Environmental Sciences of Ukraine, Kyiv, Ukraine

References

[1] AEO2011 Early Release Overview, available on the sight: www.eia.gov/forecast/aeo/pdf/0383er(2011).pdf

[2] Sharvin E.A., Aristova Ye. Yu., Syn-gas generator for internal combustion engines, "Alternative energy and ecology" (in Russain), 8 (88) 2010, c.31-37

[3] Nedybaliuk O.A., Chernyak V.Ya., Olszewskij S.V., Plasma-liquid system with reverse vortex flow of "tornado" type (Tornado-LE) //Problems of atomic science and technology, № 6. Series: Plasma Physics (16), p. 135-137. (2010).

[4] Xumei Tao, Meigui Bai, Xiang Li, e.a., CH_4-CO_2 reforming by plasma – challenges and opportunities // Progr. in Energy and Combustion Science 37, №2, pp. 113-124, 2011

[5] V. Chernyak, Eu. Martysh, S. Olszewski, D. Levko. e.a., Ethanol Reforming in the Dynamic Plasma - Liquid Systems, Biofuel Production-Recent Developments and Prospects, Marco Aurélio dos Santos Bernardes (Ed.), (2011). ISBN: 978-953-307-478-8, InTech, P.101-136.

[6] Lukes, P.; Sunka, P.; Hoffer, P.; Stelmashuk, V.; Benes, J.; e.a.,Book of Abstracts: NATO Science Advanced Research Workshop on Plasma for bio decontamination, medicine and food security, Jasná, Slovakia, March 15–18, 2011.

[7] Sheldon R. A. C. Catalytic conversions in water and supercritical carbon dioxide from the standpoint of sustainable development (in Russian) // Rus. Chem. J., 48(2004), 74-83.

[8] N.A.Popov, V.A. Shcherbakov, e.a. Thermonuclear fusion by exploding a spherical charge (gas-dynamical thermonuclear fusion problem) // Uspekhi, 10 (2008), 1087-1094.

[9] Laux, C.O. Optical diagnostics of atmospheric pressure air plasma SPECAIR / C.O. Laux, T.G. Spence, C.H. Kruger, and R.N.Zare // Plasma Source Sci. Technol. – 2003. - Vol. 12, No. 2. - P. 125-138.

[10] Raizer Yu.P. Gas discharge physics (Springer, 1991)

[11] D. Levko, A. Shchedrin, V. Chernyak e.a., Plasma kinetics in ethanol/water/air mixture in a 'tornado'-type electrical discharge // J. Phys. D: Appl. Phys. 44 (2011) 145206 (13pp)

[12] Kushner, M.J.; Babaeva, N.Yu. Plasma production in liquids: bubble and electronic mechanism, Bulletin of the APS GES10, Paris, France, October 4–8, 2010.

Chapter 6

The Promising Fuel-Biobutanol

Hongjuan Liu, Genyu Wang and Jianan Zhang

Additional information is available at the end of the chapter

1. Introduction

In recent years, two problems roused peoples' concern. One is energy crisis caused by the depleting of petroleum fuel. The other is environmental issues such as greenhouse effect, global warming, etc. Therefore, renewable sources utilization technology and bioenergy production technology developed fast for solving such two problems. Bioethanol as one of the biofuel has been applied in automobiles with gasoline in different blending proportions (Zhou and Thomson, 2009; Yan and Lin, 2009). Biobutanol is one of the new types of biofuel. It continuously attracted the attention of researchers and industrialists because of its several distinct advantages.

1.1. Property of butanol

Butanol is a four carbon straight chained alcohol, colorless and flammable. Butanol can be mixed with ethanol, ether and other organic solvent. Butanol can be used as a solvent, in cosmetics, hydraulic fluids, detergent formulations, drugs, antibiotics, hormones and vitamins, as a chemical intermediate in the production of butyl acrylate and methacrylate, and additionally as an extract agent in the manufacture of pharmaceuticals. Butanol has a 4-carbon structure and the carbon atoms can form either a straight-chain or a branched structure, resulting in different properties. There exist different isomers, based on the location of the–OH and carbon chain structure. The different structures, properties and main applications are shown as Table 1.

Although the properties of butanol isomers are different in octane number, boiling point, viscosity, etc., the main applications are similar in some aspects, such as being used as solvents, industrial cleaners, or gasoline additives. All these butanol isomers can be produced from fossil fuels by different methods, only n-butanol, a straight-chain molecule structure can be produced from biomass.

	n-Butanol	2-Butanol	iso-Butanol	tert-Butanol
Molecular structure				
Density (g/cm3)	0. 81	0. 806	0. 802	0. 789
Boiling point(°C)	118	99. 5	108	82. 4
Melting point(°C)	-90	-115	-108	25-26
Refractive index(n20D)	1. 399	1. 3978	1. 3959	1. 3878
Flash point(°C)	35	22-27	28	11
Motor octane number	78	32	94	89
Main applications	Solvents-for paints, resins, dyes, etc. Plasticizers- improve a plastic material processes Chemical intermediate -for butyl esters or butyl ethers, etc. Cosmetics- including eye makeup, lipsticks, etc. Gasoline additive	Solvent Chemical intermediate- for butanone, etc. Industrial cleaners -paint removers Perfumes or in artificial flavors	Solvent and additive for paint Gasoline additive Industrial cleaners -paint removers Ink ingredient	Solvent Denaturant for ethanol Industrial cleaners- paint removers Gasoline additive for octane booster and oxygenate Intermediate for MTBE, ETBE, TBHP, etc.

Table 1. Structures, properties and main applications of n-butanol, 2-Butanol, iso-Butanol and tert-Butanol

1.2. Advantages of butanol as fuel

Except the use of solvent, chemical intermediate and extract agent, butanol also can be used as fuel, which attracted people's attention in recent years. Because of the good properties of high heat value, high viscosity, low volatility, high hydrophobicity, less corrosive, butanol has the potential to be a good fuel in the future. The properities of butanol and other fuels or homologues are compared as Table 2. (Freeman et al., 1988; Dean, 1992)

Fuel	Octane number	Cetane number	Evaporation heat (MJ/kg)	Combustion energy(MJ/dm³)	Flammability limits (%vol)	Saturation pressure (kPa) at 38°C
Gasoline	80-99	0-10	0. 36	32	0. 6-0. 8	31. 01
Methanol	111	3	1. 2	16	6-36. 5	31. 69
Ethanol	108	8	0. 92	19. 6	4. 3-19	13. 8
Butanol	96	25	0. 43	29. 2	1. 4-11. 2	2. 27

Table 2. Properities of butanol and other fuels

Butanol appeared the good properties compared with it's homologues such as 2-butanol, iso-butanol and tert-butanol and other fuels such as Gasoline and ethanol. Actually, when ethanol is mixed with gasoline (less than 10%), there exists some disadvantages. Firstly, the heating value of ethanol is one sixth of gasoline. The fuel consumption will increase 5% if the engine is not retrofitted. Secondly, acetic acid will be produced during the burning process of ethanol, which is corrosive to the materials of vehicle. The preservative must be added when the ethanol proportion upper than 15%. Thirdly, ethanol is hydroscopic and the liquid phase separation may be occurring with high water proportion. Furthermore, ethanol as fuel cannot be preserved easily and it is more difficult in the process of allocation, storage, transition than that of gasoline.

Compared with ethanol, butanol overcomes above disadvantages and it shows potential advantages. For example, Butanol has higher energy content and higher burning efficiency, which can be used for longer distance. The air to fuel ratio and the energy content of butanol are closer to gasoline. So, butanol can be easily mixed with gasoline in any proportion. Butanol is less volatile and explosive, has higher flash point, and lower vapor pressure, which makes it safer to handle and can be shipped through existing fuel pipelines. In addition, Butanol can be used directly or blended with gasoline or diesel without any vehicle retrofit (Durre, 2007; Pfromm et al., 2010).

Actually, the first-time synthesis of biobutanol at laboratory level was reported by Pasteur in 1861 (Durre, 1998) and the industrial synthesis of biobutanol was started during 1912–1914 by fermentation (Jones and Woods, 1986). However, before 2005, butanol was mainly used as solvent and precursor of other chemicals due to the product inhibition and low butanol productivity. To bring awareness to butanol's potential as a renewable fuel, David Ramey drove his family car from Ohio to California on 100% butanol (http://www.consumerenergyreport.com /2011/02/09/reintroducing-butanol/). And then, two giant companies DuPont and BP have declared to finance development of a modernize production plant supported by research and development. (http://biomassmagazine.com/articles/2994 /eu-approves-bp-dupont-biobutanol-venture) The economy of biobutanol production also was revaluated. The research of a continuous fermentation pilot plant operating in Austria in the 1990s introduced new technologies and proved economic feasibility with agricultural waste potatoes. (Nimcevic and Gapes, 2000).

2. Production methods of butanol

Butanol can be obtained using chemical technologies, such as Oxo-synthesis and aldol condensation. It is also possible to produce butanol in the process of fermentation by bacteria and butanol as one of the products called biobutanol. The most popular bacteria species used for fermentation is *Clostridium acetobutylicum*. Because the main products of this process containing acetone, butanol and ethanol, the fermentation is called ABE fermentation (Qureshi and Maddox, 1995).

2.1. Chemical process

Butanol can be produced by chemical synthesis. One process is Oxo-synthesis, which involves the reaction of propylene with carbon monoxide and hydrogen in the presence of cobalt or rhodium as the catalyst. The mixture of n-butyraldehyde and isobutyraldehyde are obtained and then the mixture can be hydrogenated to the corresponding n-butanol and isobutyl alcohols (Park, 1996).The reactions are as following:

$$CH_3CH-CH_2+CO+H_2 \rightarrow CH_3CH_2CH_2CHO+ (CH_3)_2 CHCHO \tag{1}$$

$$\begin{align} CH_3CH_2CH_2CHO+H_2 &\rightarrow CH_3CH_2CH_2CH_2OH \text{ (a)} \\ (CH_3)_2 CHCHO+H_2 &\rightarrow (CH_3)_2 CHCH_2OH \qquad \text{(b)} \end{align} \tag{2}$$

When using cobalt as the catalyst, the reaction processes at $10\sim20MPa$ and $130\sim160°°C$, the products ratio of n-butyraldehyde and isobutyraldehyde is 3. Rhodium as the catalyst used in industry from 1976 and the reaction processes at 0.7-3MPa and 80-120°°C.The products ratio of n-butyraldehyde and isobutyraldehyde can reach 8-16. Hydrogenaration processes by using the catalyst of nickel or copper in gaseous phase or nickel in liquid phase. Some by-products can be transferred into butanol at high temperature and high pressure that will enhance the product purity.

Another route is aldol condensation, which involves the reaction of condensation and dehydration from two molecules of acetic aldehyde. And then, the product crotonaldehyde was transformed into n-butanol by hydrogenation at 180°°C and 0.2MPa. The reaction is as following: $CH_3CH=CHCHO+2H_2 \longrightarrow CH_3CH_2CH_2CH_2OH$

Comparing the two processes, Oxo-synthesis route has the advantages of materials easily obtained, comparable moderate reaction conditions, enhanced ratio of n-butanol to isobutyl alcohol. So, Oxo-synthesis process is the main industrial route for n-butanol production. There are also some other fossil oil derived raw materials such as ethylene, propylene and triethylaluminium or carbon monoxide and hydrogen are used in butanol production (Zverlov, et al., 2006).

2.2. Biological process

Except the chemical ways, butanol can also be obtained from biological ways with the renewable resources by the microorganism through fermentation. The *Clostridia* genus is very common for butanol synthesis under anaerobic conditions, and the fermentation products are often the mixture of butanol, acetone and ethanol. A few kinds of *Clostridium* can utilize cellulose and hemicellulose with the ability of cellulolytic activities (Mitchell et al., 1997; Berezina et al. 2009).

Compared with the chemical ways for butanol production, biological ways has the distinct advantages. For example, it can utilize the renewable resources such as wheat straw, corn core, switch grass, etc. Furthermore, biological process has high product selectivity, high security, less by-products. Furthermore, the fermentation condition of butanol production is milder than that of chemical ways and the products are easier to separate. The process of biobutanol production with Lignocellulosic feedstocks is as following (Fig. 1):

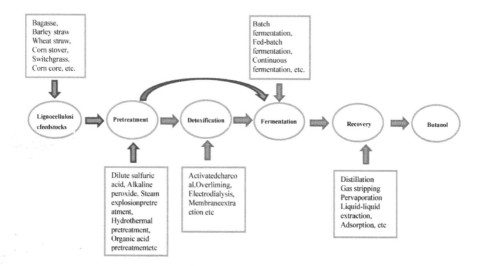

Figure 1. Butanol production process from lignocellulosic feedstocks

For the first step, biomass containing lignocellulosics should be pretreated before they were used as the substrate for the fermentation, except for a few high cellulase activity strains (Ezeji and Blaschek, 2008). The pretreatment methods are different according to the different types of biomass used. There often use dilute sulfuric acid pretreatment, alkaline peroxide pretreatment, steam explosion pretreatment, hydrothermal pretreatment, organic acid pretreatment etc. Some inhibitors such as acetic acid, furfural, 5- hydroxymethyl furfural, phenols etc. that need to be further detoxified. The ordinary detoxification methods are using activated charcoal (Wang et al., 2011), overliming (Sun and Liu, 2012; Park et al., 2010), electrodialysis (Qureshi et al., 2008c), membrane extraction (Grzenia et al., 2012) to remove the inhibitors. This step is determined by different feed stock and different pretreatment methods. After the fermentation, the desired product is recovered and purified in the downstream process. Biological ways has been set up for many years while it was inhibited for industrial application for economic reasons. So, as an alternative fuel, biomass feedstock for biobutanol production must be widely available at low cost (Kent, 2009). Therefore, by using agricultural wastes for butanol production such as straw, leaves, grass, spoiled grain and fruits etc are much more profitable from an economic point of view. Recently, other sources such as algae culture (Potts et al., 2012; Ellis et al., 2012) also is studied as one substrate for butanol production.

3. Biobutanol production by fermentation

3.1. Microbes

Clostridium is a group of obligate, Gram positive, endospore-forming anaerobes. There are lots of strains used for ABE fermentation in different culture collections, such as ATCC (American Type Culture Collection), DSM (German Collection of Microorganisms, or Deutsche Sammlung Von Mikroorganismen), NCIMB (National Collections of Industrial & Marine Bactria Ltd), and NRRL (Midwest Area National Center for Agriculture Utilization Research, US Department of Agriculture). The different strains share similar phonotype such as main metabolic pathway and end products. Molecular biology technology offers efficient method for classification. The butanol-producing clostridium can be assigned to four groups according to their genetic background, named *C. acetobutylicum, C. beijerinckii, C. saccharoperbutyl acetonicum*, and *C. saccharobutylicum*, respectively. *C. acetobutylicum* is phylogenetically distinct from the other three groups.

The common substrate for the solvent production by these strains is soluble starch. The original starch-fermenting strains belong to *C. acetobutylicum*. A recently isolated butanol-producing strain *C. saccharobutylicum* showed high hemicellulotic activity (Berezina et al., 2009). All of the four group strains can ferment glucose-containing medium to produce solvent. In 4% glucose TYA medium, *C. beijerinckii* gave the lowest solvent yield (28%), while the solvent yield was upper than 30% compared to the other three groups (Shaheen et al., 2000). In standard supplement maize medium (SMM), *C. acetobutylicum* is the best strain for maize fermentation, and the total solvent concentration can reach 19g/L. The solvent yield was 16, 14, and 11 for that of *C. beijerinckii, C. saccharoperbutyl acetonicum*, and *C. saccharobutylicum* respectively. However, *C. acetobutylicum* can't ferment molasses well and it produces bright yellow riboflavin in milk, which is different from other groups and easy identified. The best molasses-fermenting strains belong to *C. saccharobutylicum* and *C. beijerinckii* (Shaheen et al., 2000). *C. saccharoperbutyl acetonicum* can utilize sugar, molasses and maize. Comparing to *C. acetobutylicum, C. beijerinckii* was more tolerant to acetic acid and formic acid (Cho et al., 2012), which suggests the advantage when using lignocellulosic hydrolysate treated with acetic and formic acid as substrate.

There are also some *C. beijerinckii* strains produce isopropanol instead of acetone (George et al., 1983). Some microorganisms can produce biobutanol from carbon monoxide (CO) and molecular hydrogen (H_2), including acetogens, *Butyribacterium methylotrophicum, C. autoethanogenum, C. ljungdahlii* and *C. carboxidivorans*. The *C. carboxidivorans* strain P7(T) genome possessed a complete Wood-Ljungdahl pathway gene cluster which is responsible for CO, hydrogen fixation and conversion to acetyl-CoA(Fig.2) (Bruant et al., 2010).

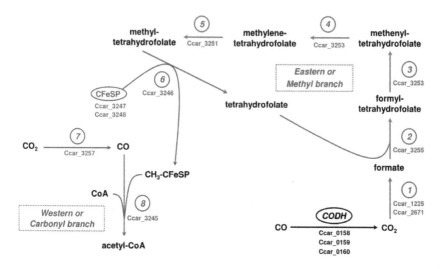

Figure 2. Wood-Ljungdahl pathway in Carboxdivorans Strain P7T. (Bruant et al. 2010, http://creativecommons. org/ licenses/by/3. 0/)Wood-Ljungdahl pathway key enzymes and protein identified in C. carboxidivorans strain P7T. 1, formate dehydrogenase; 2, formate-tetrahydrofolate ligase; 3 and 4, bifunctionalmethenyl-tetrahydrofolatecyclohydro-lase/methylene-tetrahydrofolate dehydrogenase (NADP+); 5, 5, 10-methylene-tetrahydrofolate reductase; 6, 5-methyl-tetrahydrofolate:- corrinoid iron-sulfur protein methyltransferase; 7, carbon monoxide dehydrogenase; 8, acetyl-CoA synthase; CFeSP, corrinoid iron-sulfur protein; CODH, additional carbon monoxide dehydrogenase complex. Reactions from the western branch are indicated in blue, those from the eastern branch are indicated in red. The corresponding genes in strain P7T genome are indicated below the enzyme.

3.2. Metabolic pathway

The ABE producing strains can hydrolyze starch to glucose or other hexose by amylases. Glucose was firstly converted to pyruvate through the Embden-Meyerhoff pathway (EMP, or glycolysis). Pyruvate was then cleaved to acetyl-CoA by pyruvate ferredoxin oxidoreductase. Acetyl-CoA is the common precursor of all the fermentation intermediate and end products. The enzyme activity and the coding genes have been widely assayed and described in butanol-producing strains (Dürre et al., 1995; Gheshlaghi et al., 2009).

The ABE fermentation process can be divided into two successive and distinct phase as acidogenesis phase and solvetogenesis phase. The acidogenesis phase is accompanied with cell exponential growth and pH drop, accumulation of acetate and butyrate. Solventogenesis phase begins with endospore forming and the cells entering stationary state. The products of acidogenesis phase include acetate and butyrate. Acetate forms from Acetyl-CoA, which is catalyzed by two enzymes, phosphotransacetylase (PTA, or phosphate acetyltransferase, endoced by *pta* gene) and acetate kinase (AK, encoded by *ak* gene). The butyrate synthesis is a little complicated with more steps. At first, two molecular of acetyl-CoA is catalyzed by thiolase (thl, or acetyl-CoA acetyltransferase, encoded by *thl* gene) and transforms into one molecular C4 unit acetoacetyl-CoA, which is another important node and precursor of butyrate, acetone, and butanol synthesis. The acetoacetyl-CoA is subjected to three enzymes in turn and another C4 unit butyryl-CoA is the intermediate product. The three enzymes are hydroxybutyryl-CoA dehydrogenase (encoded by *hbd* gene) (Youngleson et al., 1995), crotonase (CRT, or hydroxybutyryl-CoA dehydrolase, encoded by *crt* gene), and butyryl-CoA dehydrogenase (BCD, encoded by *bcd* gene). Accordingly, three encoded genes coexist in the BCS operon with additional two genes coding for the α and β subunit of electron transfer protein (Bennett and Rudolph, 1995). Butyryl-CoA was then catalyzed by phosphotransbutylase (PTB, or phosphate butyltransferase, encoded by *ptb* gene) and butyrate kinase (BK, encoded by *bk* gene) to form butyrate during acidogenesis phase.

As the organic acid accumulation, pH drop to the lowest point during the fermentation. This leads to the switch of acidogenesis phase to solventogenesis phase. Acetate and butyrate are reassimilated and participate in the solvent formation. Under the catalyzing of CoA transferase (CoAT, two unit encoded by *ctfα* and *ctfβ*), acetate and butyrate was transformed into acetyl-CoA and butyryl-CoA respectively again. The alcohols formation share the same key enzymes, NAD(P)H dependent aldehyde/alcohol dehydrogenases (encoded by *adh*1 and *adh*2 gene) (Chen, 1995). In addition, Butanol owns its unique butanol dehydrogenase (encoded by *bdh* gene) (Welch et al., 1989). The formation of acetone from acetoacetyl-CoA is a two-step reaction. Acetoacetyl-CoA is catalyzed to acetoacetate by CoA transferase. Acetone is produced after a molecular CO_2 released from acetoacetate by decarboxylase (AADC, encoded by *aadc* gene) (Janati-Idrissi et al., 1988; Cary et al., 1993). Both acid reassimilation and acetone formation utilize CoA transferase, however, the butyrate uptake was not concomitant with the production of acetone (Desai et al., 1999). The metabolic pathway accompanied by electron transfer and reduction force forming. The main ABE fermentation pathway was illustrated in Fig.3.

Solventogenic genes *aad*, *ctfA*, *ctfB* and *adc* constitute the *sol* operon (Durre et al., 1995). In some conditions, butanol producing strains lose the ability to produce solvents after repeated subculturing, called as degenerated (DGN) strain. In *C. acetobutylicum* ATCC 824, the plasmid pSOL1 carrying the *sol* operon was found missing during degenerating process (Cornillot et al., 1997). For *C. saccharoperbutyl acetonicum* strain N1-4, the *sol* genes maintained in degenerated DGN3-4 strain, while the *sol* operon was hardly induced during solventogenesis. Extract from the culture supernatants of wild-type N1-4 is enough to induce the transcription of the *sol* operon in DGN3-4 (Kosaka et al., 2007). It suggested that the degeneration maybe caused by the incompetence of the induction mechanism of the *sol* operon. The transcription of *sol* operon may be under the control of the quorum-sensing mechanism in *C. saccharoperbutyl acetonicum*.

Though the metabolic pathway is clear, the underlying regulation mechanism is poorly understood, such as the phase switch of fermentation, the relationship between solventogenesis and sporulation. Answering these questions is critical to improve the efficiency of butanol producing fundamentally. Proteomics and transcriptomics can provide more unknown details, which will be helpful for solving these problems (Sivagnanam et al., 2011; Sivagnanam et al., 2012).

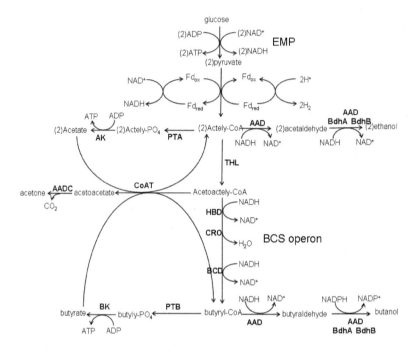

Figure 3. Metabolic pathway of Acetone-butanol-ethanol fermentation. EMP: Embden-Meyerhoff pathway (glycolysis); AK, acetate kinase; PTA, phosphotransacetylase; CoAT, CoA transferase; AADC, acetoacetate decarboxylase; THL, thiolase; BK, butyrate kinase; PTB, phosphotransbutylase; HBD, hydroxybutyryl-CoA dehydrogenase; CRO, crotonase; BCD, butyry-CoA dehydrogenase; AAD, aldylde/ alcohol dehydrogenase; BdhA, butyryl-CoA dehydrogenase A; BdhB, butyryl-CoA dehydrogenase B.

3.3. Metabolic engineering

The increasing genetic knowledge provides feasible technique for the strain modification. Many efforts have been made to construct the strain with high butanol tolerance, superior butanol yield, productivity and less byproduct. The process can be classified into pathway-based construction and regulation-based construction.

Except butanol, acetone and ethanol are main products in ABE fermentation. The byproduct, especially acetone is low valuable and undesirable. Blocking the expression key enzyme gene for acetone is thought perfect to decrease the split flux and enhance butanol yield. However, the results were not ideal as expected. Knocking out the C. acetobutylicum EA 2018 *adc* gene, the acetone is still produced in low level (Jiang et al., 2009). In *C. beijerinckii* 8052, the strain with *adc* gene disruption produced similar acetone with the original wild type strain (Han et al., 2011). To block acetate and acetone pathway by knocking out gene *adc* and *ctfA* reduced solvent production (Lehmann et al., 2012). These results demonstrated that the butanol metabolic mechanism is more complicated than expected.

Acetate and butyrate are produced during acidogenesis, and then they are transformed into acetyl-CoA and butyl-CoA to participate the solvent formation during solventogenesis phase. It seems an ineffective loop. In fact, the "inefficiency" loop is necessary for acid accumulation and switching to solventogenesis, at the same time, energy and reduction force were reserved. Disruption of acetate and butyrate pathway didn't enhance butanol production. Knocking out acetate biosynthetic pathway gene by Clos Tron had no significant influence on the metabolite distribution (Lehmann et al., 2012). Disruption of *ptb* gene blocked the butyrate synthesis and led to acetate and lactate accumulation. Some mutant strain without *bk* gene even can't survival (Sillers et al., 2008). It indicated that the pathways seeming useless were necessary for butanol synthesis. What's more, it is not possible to improve performance by decrease acid formation.

The genes participate in butanol synthesis including of *thl*, BCS operon, and *add*, *bdh*. Overexpression these genes are thought useful to increase the butanol yield. Overexpression of *aad* gene alone could enhance butanol production (Nair and Papoutsakis, 1994; Tummala et al., 2003). Transformed strain M5 (*sol* operon deficient because of lose of plasmid pSOL) with a plasmid carrying *aad* gene restored butanol-producing capability (Nair and Papoutsakis, 1994). Overexpression of *aad* gene and down-regulated *ctf* gene increased the butanol and ethanol production. To boost the butyryl-CoA pool, the strain with both *thl* and *aad* overexpression was constructed. However, butyrate and acetone concentration were increased, not butanol. The *thl* overexpression with *ctf* knock down didn't change the product significantly (Sillers et al., 2009). So, the metabolic is more complicated than it seems. Theoretical analyses also suggested alteration single solvent-associated gene is not sufficient to increase butanol yield (Haus et al., 2011).

Low butanol tolerance of the strains is another problem of butanol production. Although butanol synthesis is spontaneous in clostridium, the wild type strains can't endure high butanol concentration upper than 2%. Butanol stress influence gene expression of amino acid, nucleotide, glycerolipid biosynthesis and the cytoplasmic membrane composition (Janssen et al., 2012). Cells have heat shock response system will protect it from heat or other stress (Bahl, Müller et al. 1995). Overexpression of grosESL improved the strain tolerance and butanol titer (Tomas et al., 2003).

The utilization of xylose and other carbon sources was inhibited by glucose is a phenomenon called as Carbon catabolite repression (CCR). CCR limited the efficiency of butanol fermentation with lignocellulosic material as substrate. The utilization rate of pentose was improved efficiently by knocking out pleiotropic regulator gene *ccpA*, *glcG* (responsibility for phosphoenoopyruvate-dependent phophotransferase system, PTS) and overexpressing the genes of xylose utilization (Ren et al., 2010; Xiao et al., 2012). By heterogonous expression transaldolase gene talA in ATCC 824, the xylose utilization was improved significantly (Gu et al., 2009). Knocking out xylose repressor gene XylR also increased the fermentation efficiency (Xiao et al., 2012).

There also some strategies aim at the upstream regulation. Global transcription machinery engineering (gTME) is thought to be a promising method to improve the butanol-producing performance (Alper et al., 2006; Papoutsakis, 2008). By regulating the transcription factor, the gTME strategy is thought to be able to change the metabolic strength and direction. gTME has been shown an efficient solution to improve substrate utilization, product toler-ance, and production in yeast (Alper et al. 2006) and *E. coli* (Chen et al., 2011). In butanol-producing *Clostridium*, the metabolic pathway have been described clearly, however, the mechanism of metabolism regulation is still not fully understood. This situation keeps the gTME strategy away from butanol-producing strains. Much effort should be devoted on the proteomics and transcriptomics etc. that will increase more details behind the appearance of ABE fermentation. A true gTME strategy will bring fresh and effective innovation to the bu-tanol fermentation.

The concept of metabolic engineering is to develop strains as "cell factory" which is efficient for desired products production from renewable sources (Na et al., 2010). Some microbes at-tracted interests because they are more tolerant to butanol than *Clostridium*, although these bacteria haven't natural solvent-producing ability. Some kinds of Lactic acid bacteria can grow in 3-4% butanol (Liu et al., 2012) after long term adaption, that makes them promising host for butanol producing. The synthetic biology strategy has been implemented by con-structing the whole butanol-producing pathway in *Escherichia coli*, *Bacillus subtilis*, *Saccharo-myces cerevisiae* and *Pseudomonas putida* (Shen and Liao, 2008; Nielsen et al., 2009). This strategy deserves further attempts in spite of the poor final butanol concentration.

3.4. Fermentation application

ABE fermentation can be conducted as batch, fed-batch, and continuous under anaerobic conditions. Batch fermentation is the simplest mode. The substrate is typical 40-80g/L and the efficiency decreased as substrate concentration upper than 80g/L (Shaheen et al, 2000). With optimized physiological and nutritional parameters, 20g/L n-butanol was obtained by *C. beijerinckii* ATCC 10132 in 72h (Isar and Rangaswamy, 2012). Fed-batch fermentation was adopted to avoid substrate inhibition. However, because of product inhibition, the substrate feeding seems ineffective. The solvent must be removed from the broth to decrease the product toxicity. The solvent can be removed by several ways such as liquid-liquid extrac-tion, perstraction, gas-stripping, and pervaporation etc. (Qureshi and Maddox, 1995; Qure-shi and Blaschek, 2001b). The whole systemic technique of high productivity was constructed by continuous feeding combined with product removal (Qureshi et al., 1992), such as using membrane reactor (Qureshi et al., 1999a). With these techniques, the fermenta-tion can be continuing for a long time and resulting in higher productivity. To improve the utilization efficiency of cells, the immobilization system is used (Huang et al., 2004; Qureshi et al., 2000; Lienhardt et al., 2002). Comparing with the free cell system, the immobilization system is easier to separate cells from product, can reach high cell concentration and pro-ductivity, and can decrease nutrient depletion and product inhibition.

Co-culture is another important way for butanol fermentation (Abd-Alla and El-Enany, 2012). *C. beijerinckii* NCIMB 8052 was entangled with ATCC 824 and thought as *C. acetobuty-*

licum before the 16S rDNA based method was exploited (Johnson and Chen, 1995). These data implied that they could be cocultured before isolation. A microflora of four strain isolated from hydrogen-forming sludge of sewage performed a little high solvent yield (Cheng et al., 2012). Different strains possess various advantages, either with larger carbon substrate, higher butanol yield, or with high substrate and product tolerance. The co-culture should possess potential benefits and be harnessed fully after all the details are disclosed for each individual strain.

4. Separation of butanol product

Because butanol has a higher boiling point than water, therefore, distillation is not suitable for butanol recovery. Other processes such as adsorption, pervaporation, membrane pertraction, reverse osmosis and gas stripping have been developed to improve recovery performance and reduce costs (Oudshoorn et al., 2009; Ezeji et al., 2004b).

4.1. Adsorption process

Adsorption is the technology operating easily for the butanol separation. Butanol can be adsorpted by the adsorbents in the fermenter and then the butanol was obtained by desorption. A variety of materials can be used as adsorbents for butanol recovery and silicalite is the common one used (Qureshi et al., 2005b; Ezeji et al., 2007). Silicalite is a form of silica with a zeolite-like structure and hydrophobic properties, it can selectively adsorb small organic molecule like C1–C5 alcohols from dilute aqueous solutions (Zheng et al., 2009). However, adsorption separation process is not suitable on an industrial or semi-technical scale because the capacity of adsorbent is very low.

4.2. Butanol recovery by membrane reactor

Immobilization of microorganisms in the membrane or using membrane reactors is another option of butanol removal. The productivity can be enhanced obviously by this way. Huang et al. reported the continuous ABE fermentation by immobilized *C. acetobutylicum* cells with the fibrous as carrier and a productivity of 4.6 g/L/h was obtained (Huang et al., 2004). Qureshi et al. studied the butanol fermentation by immobilized *C. beijerinckii* cells with different carriers such as clay brick, the reactor productivity was enhanced to 15.8 g/(lh) (Qureshi and Blaschek, 2005a). Although the butanol productivity increased by using immobilized cell fermentation, leakage of cells from the matrices is a frequent problem for the industrial application. There still some other problems such as poor mechanical strength and increase mass transfer resistance etc.

4.3. Butanol recovery by gas stripping

Gas stripping seems to be a promising technique that can be applied to butanol recovery combined with ABE fermentation. When the gas (ordinary N_2 or CO_2) are bubbled through the fermentation broth, it captures the solvents. The solvents then condensed in the condenser and are collected in a receiver. Ezeji applied gas stripping on the fed-batch fermentation, 500 g glucose was consumed and 233 g/l solvent was produced with the productivity of 1.16 g/(Lh) and the yield of 0.47 g/g.When combined with continuous fermentation with gas stripping, 460g/l solvent was obtained with 1163g glucose consuming (Ezeji et al., 2004a; Ezeji et al., 2004b).

4.4. Butanol recovery by pervaporation

Pervaporation is a membrane-based process that allows selective removal of volatile compounds from fermentation broth. The membrane is placed in contact with the fermentation broth and the volatile liquids or solvents diffuse through the membrane as a vapor which is recovered by condensation. A vacuum applied to the side of permeate. Polydimethylsiloxane membranes and silicon rubber sheets are generally used for the pervaporation process. Selection of a suitable polymer forming the active part of the membrane is a key factor in this case. In the batch fermentation, Evans and Wang increased the solvent concentration and productivity from 24.2g/l and 0.34g/(lh) to 32.8g/l and 0.5g/(lh) with pervaporation (Evans and Wang, 1988). Groot et al. applied pervaporation on the fed-batch fermentation and the solvent productivity and concentration reached 0.98g/lh and 165.1g/l (Groot et al., 1984). The Reverse osmosis is another recovery technique that based on membranes. Before the reverse osmosis is carried out, the suspended vegetative organisms must be removed using the hollow-fiber ultra-filter. After the pretreatment, reverse osmosis starts to dewater the fermentation liquor by rejecting solvents but allowing water to pass through the membrane. And then, the products are concentrated (Zheng et al., 2009).

4.5. Liquid–liquid extraction

Liquid–liquid extraction can be used to remove solvents from the fermentation broth. In this process, the water-insoluble organic extractant is mixed with the fermentation broth. Butanol is more soluble in the organic (extractant) phase than in the aqueous (fermentation broth) phase. So, butanol can be selectively concentrated in the organic phase. As the extractant and fermentation broth are immiscible, the extractant can easily be separated from the fermentation broth after butanol extraction. (Qureshi and Blaschek, 1999a). However, there still some problems with liquid–liquid extraction such as toxicity of extractant, extraction solvent losing, the formation of an emulsion, etc. Oleyl alcohol as a good extractant with relatively low-toxic has been used widely by the researchers (Karcher et al., 2005; Ezeji, 2006).

4.6. Application of ionic liquids

The butanol extraction process using conventional solvents may be useful, but the solvents used are often volatile, toxic and dangerous. In recent years, a growing interest in ionic liquids(IL) which also can be used in butanol recovery. Ionic liquids are organic salts present in the liquid state at room conditions, have very low vapor pressure and low solubility in water. Hence, Ionic liquids is valuable solvent in the extraction process from aqueous solutions (Fadeev and Meagher, 2001; Garcia-Chavez et al., 2012). Ionic liquids as the non-volatile, environment friendly solvents have been used in various chemical processes. With the development of the technology, ionic liquids extraction would be more promising for butanol recovery.

5. Biobutanol production from renewable resources

Biobutanol is no doubt a superior candidate renewable energy facing the exhausted fossil-energy. The clostridium can incorporate simple and complex soluble sugar, such as corn, molasses, cassava, and sugar beet. The ABE fermentation is also a solution to deal with agriculture residue, spoilage material, and domestic organic waste (Table 3). Additionally, using renewable resources is also ideal for environment problem solving.

Raw materials	Bacterial strain	Fermentation process	ABE concentration (g/l)	ABE Yield (g/g)	ABE productivity (g/lh)	References
Barley straw	C. beijerinckii	Batch fermentation	26.64	0.43	0.39 g/lh	Qureshi et al., 2010a
Wheat straw	C.beijerinckii	simultaneous saccharification and fermentation combined with gas stripping	21.42	0.41	0.31	Qureshi et al., 2008a
Corn fiber	C. beijerinckii	Batch fermentation	9.3	0.39	0.10	Qureshi et al., 2008b
Corn stover	C. beijerinckii	Batch fermentation	26.27	0.44	0.31	Qureshi et al., 2010b
Rice straw	C. saccharoperbutylacetonicum	Batch fermentation	13	0.28	0.15	Soni et al.(1982)
Bagasse	C. saccharoperbutylacetonicum	Batch fermentation	18.1	0.33	0.3	Soni et al.(1982)
Switch grass (Panicum virgatum)	C. beijerinckii	Batch fermentation	14.61	0.39	0.17	Qureshi et al., 2010a
Domestic organic waste	C. acetobutylocum	Batch fermentation	9.3	0.38	0.08	Claassen et al., 2000
Sago	C. saccharobutylicum	Batch fermentation Continuous Fermentation (D=0.11h⁻¹)	16.38 7.74±0.55	0.33 0.29	0.59 0.85	Liew et al., 2005
Defibrated-sweet potato-slurry (DSPS)	C. acetobutylocum	Batch fermentation Continuous Fermentation, immobilized cell (D=0.129 h⁻¹)	5.87 7.73	0.29 0.195	0.12 1	Badr et al., 2001
Cassava	Co-culture of B. Subtilis and C. butylicum	Batch fermentation	9.71	~0.21	0.135	Tran et al., 2010

Table 3. Butanol production with different raw materials

Food-based substrate arouses many problems. The cost of butanol from glucose was four fold higher than that from sugarcane and cellulose materials (Kumar et al., 2012). For the cellulose-based substrate, the crystal structure of cellulose is hard to use for normal ABE fermentation clostridium. The pretreatment of cellulose is costly, complex, and often leads to new environment problems. For example, using corn as substrate, the cost is 0.44-0.55 US $/kg butanol by the hyper-butanol producing strain *C. beijerinckii* BA101 (Qureshi and Blaschek, 2000) by continuous fermentation combined with butanol separation. The cost reached 0.73-1.07 US$/kg when grass-rooted plant was used as substrate (Qureshi and Bla-schek, 2001a). A promising solution is co-culture of butanol-producing and cellulolytic strains. However, many obstacles must be cleared before the system is constructed. It's diffi-cult for different strains to play a role in turn in the substrate medium. Firstly, strain with high hydrolysis activity must be obtained. Secondly, the procedure must also be optimized.

Some strains can use CO_2, H_2, and CO as substrate (Tracy et al., 2012). The celluloses sub-strate can be transformed into CO $(_2)$ and H_2 firstly. The simple substrates then are used by *C. carboxidivorans* to produce butanol. The more simple and feasible process is still need to be further explored for different substrates.

6. The promising application and prospect of biobutanol

Due to the excessive exploitation, the fossil fuels are facing scarce and they cannot be generated. On the other hand, most of the carbon emissions result from fossil fuel combustion. Reducing the use of fossil fuels will considerably reduce the amount of carbon dioxide and other pollutants produced. Renewable energy has the potential to provide energy services with low emissions of both air pollutants and greenhouse gases. Currently, renewable energy sources supply over 14% of the total world energy demand. Biofuels as the important renewable energy are generally considered as sustainability, reduction of greenhouse gas emissions, regional development, so-cial structure and agriculture, and security of supply (Reijnders, 2006). Biodiesel and bioethanol are presently produced as a fuel on an industrial scale, including ETBE partially made with bioe-thanol, these fuels make up most of the biofuel market (Antoni et al., 2007).

Biobutanol also has a promising future for the excellent fuel properties. It has been demon-strated that n-butanol can be used either 100% in unmodified 4-cycle ignition engines or blended up with diesel to at least 30% in a diesel compression engine or blended up with kerosene to 20% in a jet turbine engine in 2006 (Schwarz et al., 2006). The production of bio-butanol from lignocellulosic biomass is promising and has been paid attention by many companies. Dupont and BP announced a partnership to develop the next generation of bio-fuels, with biobutanol as first product (Cascone, 2007). In 2011, Cobalt Technologies Compa-ny and American Process Inc. (API) have been partnering to build an industrial-scale cellulosic biorefinery to produce biobutanol. Additionally, the companies agreed to jointly market a GreenPower+ biobutanol solution to biomass power facilities and other customers worldwide. The facility is expected to start ethanol production in early 2012 and switch to biobutanol in mid-2012. The annual production of biobutanol is estimated to 470, 000 gal-lons. (http://www.greencarcongress.com/2011/04/cobalt-20110419.html, http://www.renewa-bleenergyfocususa.com/view/17558/cobalt-and-api-cooperate-on-biobutanol/) Gevo, Inc. signed a Joint Development Agreement with Beta Renewables, a joint venture between Chemtex and TPG, to develop an integrated process for the production of bio-based isobuta-nol from cellulosic, non-food biomass, such as switch grass, miscanthus, agriculture resi-dues and other biomass will be readily available. (http://www.greencarcongress.com/biobutanol/). Syntec company also is currently developing catalysts to produce bio-butanol from a range of waste biomass, including Municiple Solid Waste, agricultural and forestry wastes. (http://www.syntecbiofuel.com/butanol.php). Utilization the waste materials improve the economy of butanol production that makes biobutanol great potential to be the next new type of biofuel in spite of the existing drawbacks.

7. Conclusions

Biobutanol production has only recent years booming again after long time of silence. Quite a lot of progress has been made with the technology development of metabolic engineering in enhancing solvent production, increasing the solvent tolerance of bacteria, improving the selectivity for butanol. Fortunately, *Clostridia* have been tested being able to consume lignocellulosic biomass for ABE fermentation. The complex regulation mechanism of butanol synthesis is still need to be further study. For the strain improvement, for example, constructing better butanol tolerance strains, more suitable hosts and genetic methods are required to be set up. Furthermore, more efficient techniques for removing the inhibitors in the lignocellulosic hydrolysate need to be developed. In addition, from the economic point of view, the integrated system of hydrolysis, fermentation, and recovery process also are important to be further developed to reduce the operation cost of butanol synthesis.

Author details

Hongjuan Liu*, Genyu Wang and Jianan Zhang

*Address all correspondence to: liuhongjuan@tsinghua.edu.cn; zhangja@tsinghua. edu. cn

Institute of Nuclear and New Energy Technology, Tsinghua University, Beijing, P. R., China

References

[1] Abd-Alla MH, El-Enany AWE. Production of acetone-butanol-ethanol from spoilage date palm (Phoenix dactylifera L.) fruits by mixed culture of Clostridium acetobutylicum and Bacillus subtilis. Biomass Bioenergy. 2012, 42: 172-178.

[2] Alper H, Moxley J, Nevoigt E, Fink GR, Stephanopoulos G. Engineering yeast transcription machinery for improved ethanol tolerance and production. Science. 2006, 314(5805): 1565-1568.

[3] Antoni D, Zverlov VV, Schwarz WH. Biofuels from microbes. ApplMicrobiolBiotechnol. 2007, 77:23–35

[4] Badr HR, Toledo R, Hamdy MK. Continuous acetone ethanol butanol fermentation by immobilized cells of Clostridium acetobutylicum. Biomass Bioenergy. 2001, 20:119–132

[5] Bennett GN, Rudolph FB. The central metabolic pathway from acetyl-CoA to butyryl-CoA in Clostridium acetobutylicum. FEMS Microb Rev. 1995, 17(3): 241-249.

[6] Berezina OV, Brandt A, Yarotsky S, Schwarz WH, ZverlovVV. Isolation of a new butanol-producing Clostridium strain: High level of hemicellulosic activity and structure of solventogenesis genes of a new Clostridium saccharobutylicum isolate. SystApplMicrobiol. 2009, 32(7): 449-459.

[7] Bruant G, Levesque MJ, Peter C, Guiot SR, Masson L. Genomic Analysis of Carbon Monoxide Utilization and Butanol Production by Clostridium carboxidivorans Strain P7(T). Plos One. 2010, 5(9).

[8] Cary JW, Petersen DJ, Papoutsakis ET, Bennett GN. Sequence and arrangement of genes encoding enzymes of the acetone-production pathway of Clostridium acetobutylicum ATCC 824. Gene. 1993, 123(1): 93-97.

[9] Cascone, R. Biofuels: What is beyond ethanol and biodiesel? Hydrocarbon. 2007, 86(9)95-109.

[10] Chen JS. Alcohol dehydrogenase: multiplicity and relatedness in the solvent-producing clostridia. FEMS Microb Rev. 1995, 17(3): 263-273.

[11] Chen T, Wang J, Yang R, Li J, Lin M, Lin Z. Laboratory-evolved mutants of an exogenous global regulator, IrrE from Deinococcus radiodurans, enhance stress tolerances of Escherichia coli. PLoS One. 2011, 6(1): e16228.

[12] Cho, DH, Shin SJ, Kim YH. Effects of acetic and formic acid on ABE production by Clostridium acetobutylicum and Clostridium beijerinckii. BiotechnolBioproc E. 2012, 17(2): 270-275.

[13] ClaassenPAM, BuddeMAW, López-Contreras AM. Acetone, butanol and ethanol production from domestic organic waste by solventogenic clostridia. J MolMicrob Biotech. 2000, 2(1): 39-44.

[14] CornillotE, NairRV, Papoutsakis ET, Soucaille P. The genes for butanol and acetone formation in Clostridium acetobutylicum ATCC 824 reside on a large plasmid whose loss leads to degeneration of the strain. J Bacteriol. 1997, 179(17): 5442-5447.

[15] Dean JA. Lange's handbook of chemistry. 14th edition. New York: McGraw-Hill;1992

[16] Desai RP, Harris LM, Welker NE, Papoutsakis ET. Metabolic Flux Analysis Elucidates the Importance of the Acid-Formation Pathways in Regulating Solvent Production by Clostridium acetobutylicum. Metablic Eng. 1999, 1(3): 206-213.

[17] Dürre P, Fischer RJ, Kuhn A, Lorenz K, Schreiber W, Stürzenhofecker B, Ullmann S, Winzer K, Sauer U. Solventogenic enzymes of Clostridium acetobutylicum: catalytic

properties, genetic organization, and transcriptional regulation. " FEMS Microb Rev. 1995, 17(3): 251-262.

[18] Dürre P. Biobutanol: an attractive biofuel. Biotechnol J. 2007, 2:1525–1534.

[19] Dürre P. New insights and novel developments in clostridial acetone/butanol/isopropanefermentation. ApplMicrobBiotechnol, 1998, 49:639–648.

[20] Ellis JT, Hengge NN, Sims RC, Miller CD. Acetone, butanol, and ethanol production from wastewater algae. Bioresource Technol. 2012, 111:491-495.

[21] Evans PJ, Wang HY. Enhancement of butanol formation by Clostridium acetobutylicum in the presence of decanol-oleyl alcohol mixed extractants. Appl Environ Microbiol. 1988, 54:1662–1667.

[22] Ezeji T, Blaschek HP. Fermentation of dried distillers' grains and solubles (DDGS) hydrolysates to solvents and value-added products by solventogenic clostridia. Bioresource Technol. 2008, 99(12): 5232-5242.

[23] Ezeji TC, Qureshi N, Blaschek HP. Acetone butanol ethanol (ABE) production from concentrated substrate: reduction in substrate inhibition by fed-batch technique and product inhibition by gas stripping. ApplMicrobiolBiotechnol. 2004a, 63:653–8.

[24] Ezeji TC, Qureshi N, Blaschek HP. Bioproduction of butanol from biomass: from genes to bioreactors. CurrOpinBiotechnol, 2007, 18:220-227.

[25] Ezeji TC, Qureshi N, Blaschek HP. Butanol fermentation research: Upstream and downstream manipulations. Chem Rec. 2004b, 4:305–314.

[26] Ezeji TC, Qureshi N, Karcher P, Blaschek HP. Butanol production from corn. In Alcoholic Fuels: Fuels for Today and Tomorrow. Edited by Minteer SD. New York, NY: Taylor & Francis, 2006:99-122.

[27] Fadeev AG, Meagher MM. Opportunities for ionic liquids in recovery of biofuels. ChemCommun. 2001, 295-296.

[28] Freeman J, Williams J, Minner S, Baxter C, DeJovine J, Gibbs L, Lauck J, Muller H,. Saunders H. Alcohols and ethers: a technical assessment of their application as fuels and fuel components, API publication 4261. 2nd ed. New York: American Institute of Physics; 1988.

[29] Garcia-Chavez LY, Garsia CM, Schuur B, de Haan AB. Biobutanol Recovery Using Nonfluorinated Task-Specific Ionic Liquids. Ind Eng Chem Res. 2012, 51(24): 8293-8301.

[30] George HA, Johnson JL, Moore WE, Holdeman LV, Chen JS. Acetone, Isopropanol, and Butanol Production by Clostridium beijerinckii (syn. Clostridium butylicum) and Clostridium aurantibutyricum. Appl Environ Microbiol. 1983, 45(3): 1160-1163.

[31] GheshlaghiR, Scharer JM, Moo-Young M, Chou CP. Metabolic pathways of clostridia for producing butanol. Biotechnol Adv. 2009, 27(6): 764-781.

[32] Groot WJ, Oever CE van den, Kossen NWF. Pervaporation for simultaneous product recovery in the butanol/isobutanol batch fermentation. BiotechnolLett. 1984, 6:709–714.

[33] Grzenia DL, Schell DJ, Wickramasinghe SR. Membrane extraction for detoxification of biomass hydrolysates. Bioresource Technol. 2012, 111:248-254.

[34] Gu Y, Li J, Zhang L, Chen JNiu LX, Yang YL, Yang S, Jiang WH. Improvement of xylose utilization in Clostridium acetobutylicum via expression of the talA gene encoding transaldolase from Escherichia coli. J Biotechnol. 2009, 143(4): 284-287.

[35] Haus S, Jabbari S, Millat T, Janssen H, Fischer RJ, Bahl H, King JR, Wolkenhauer O. A systems biology approach to investigate the effect of pH-induced gene regulation on solvent production by Clostridium acetobutylicum in continuous culture. BMC Syst Biol. 2011, 5: 10.

[36] Huang WC, Ramey DE, Yang ST. Continuous production of butanol by Clostridium acetobutylicum immobilized in a fibrous bed reactor. ApplBiochemBiotechnol. 2004, 113:887-898.

[37] Isar J, Rangaswamy V. Improved n-butanol production by solvent tolerant Clostridium beijerinckii. " Biomass Bioenerg. 2012, 37: 9-15.

[38] Janati-IdrissiR, JunellesAM, Petitdemange H, Gay R. Regulation of coenzyme a transferase and acetoacetate decarboxylase activities in clostridium acetobutylicum. " Annales de l'Institut Pasteur / Microbiologie. 1988, 139(6): 683-688.

[39] Janssen H, Grimmler C, Ehrenreich A, Bahl H, Fischer RJ. A transcriptional study of acidogenic chemostat cells of Clostridium acetobutylicum—Solvent stress caused by a transient n-butanol pulse. J Biotec. http://dx. doi. org/10. 1016/j. jbiotec. 2012. 03. 027.

[40] Jiang Y, Xu CM, Dong F, Yang YL, Jiang WH, Yang S. Disruption of the acetoacetate decarboxylase gene in solvent-producing Clostridium acetobutylicum increases the butanol ratio. Metab Eng. 2009, 11(4–5): 284-291.

[41] Johnson JL, Chen JS. Taxonomic relationships among strains of clostridium-acetobutylicum and other phenotypically similar organisms. FEMS Microbiol Rev. 1995, 17(3): 233-240.

[42] Jones DT, Woods DR. Acetone-Butanol fermentation revisited. Microbiol Rev 1986, 50(4):484–524.

[43] Karcher P, Ezeji TC, Qureshi N, Blaschek HP. Microbial production of butanol: product recovery by extraction. In Microbial Diversity: Current Perspectives and Potential Applications. Edited by Satyanarayana T, Johri BN. New Delhi: IK International Publishing House Pvt. Ltd; 2005, 865-880.

[44] Kent SK. Biofuels in the U. S. —challenges and opportunities. Renew Energy 2009, 34:14–22.

[45] Kosaka T, Hirakawa H, Matsuura K, Yoshino S, Furukawa K. Characterization of the sol operon in butanol-hyperproducing Clostridium saccharoperbutylacetonicum strain N1-4 and its degeneration mechanism. Biosci Biotech Bioch. 2007, 71(1): 58-68.

[46] Kumar M, Goyal Y, Sarkar A, Gayen K. Comparative economic assessment of ABE fermentation based on cellulosic and non-cellulosic feedstocks. Appl Energy. 2012, 93: 193-204.

[47] Lehmann D, Hönicke D, Ehrenreich A, Schmidt M, Weuster-Botz D, Bahl H. Modifying the product pattern of Clostridium acetobutylicum: physiological effects of disrupting the acetate and acetone formation pathways. Appl Microbiol Biotechnol. 2012, 94(3): 743-754.

[48] Lienhardt J, Schripsema J, Qureshi N, BlaschekHP. Butanol production by Clostridium beijerinckii BA101 in an immobilized cell biofilm reactor - Increase in sugar utilization. Appl Biochem Biotechnol. 2002, 98: 591-598.

[49] Liew ST, Arbakariya A, Rosfarizan M, Raha AR. Production of solvent (acetonebutanol- ethanol) in continuous fermentation by Clostridium saccharobutylicum DSM 13864 using gelatinised sago starch as a carbon source. Malays J Microbiol. 2005, 2(2): 42–45

[50] Liu S, Wilkinson BJ, Bischoff KM, Hughes SR, Rich JO, Cotta MA. Adaptation of lactic acid bacteria to butanol. "BiocatalAgriBiotechnol. 2012, 1(1): 57-61.

[51] Mitchell WJ. Physiology of Carbohydrate to Solvent Conversion by Clostridia. Adv-Microb Physiol. R. K. Poole, Academic Press. 1997, 39: 31-130.

[52] Na D, Kim TY, Lee SY. Construction and optimization of synthetic pathways in metabolic engineering. CurrOpinMicrobiol. 2010, 13(3): 363-370.

[53] Nair RV, Papoutsakis ET. Expression of plasmid-encoded aad in Clostridium acetobutylicum M5 restores vigorous butanol production. J Bacteriol. 1994, 176(18): 5843-5846.

[54] NielsenDR, LeonardE, Yoon SH, Tseng HC, Yuan CJ, Prather KJ. Engineering alternative butanol production platforms in heterologous bacteria. Metab Eng. 2009, 11(4–5): 262-273.

[55] Nimcevic D, Gapes JR. The acetone–butanol fermentation in pilot plant and pre-industrial scale. JMolMicrobiolBiotechnol. 2000, 2:15–20.

[56] Oudshoorn A, Van der Wielen LAM, Straathof AJJ. Assessment of options for selective 1-butanol recovery from aqueous solution. IndEngChem Res, 2009, 48:7325-7336.

[57] Papoutsakis ET. Engineering solventogenicClostridia. CurrOpin Biotech. 2008, 19(5): 420-429.

[58] Park CH. Pervaporativebutanol fermentation using a new bacterial strain. Biotechnol Bioprocess Eng 1996, 1:1–8.

[59] Park J, Shiroma R, Al-Haq MI, Zhang Y, Ike M, Arai-Sanoh Y, Ida A, Kondo M, To-kuyasu K. A novel lime pretreatment for subsequent bioethanol production from rice straw – Calcium capturing by carbonation (CaCCO) process. Bioresour. Technol. 2010, 101(17): 6805-6011.

[60] Pfromm PH, Boadu VA, Nelson R, Vadlani P, Madl R. Bio-butanol vs. bio-ethanol: a technical and economic assessment for corn and switch grass fermented by yeast or Clostridium acetobutylicum. Biomass Bioenerg. 2010, 34(4):515-524.

[61] Potts T, Du JJ, Paul M, May P, Beitle R, Hestekin J. The production of butanol from Jamaica bay macro algae. Environ Prog Sustain Energy. 2012, 31(1):29-36.

[62] Qureshi N, Blaschek HP. ABE production from corn: a recent economic evaluation. " J IndMicrobiolBiot. 2001a, 27(5): 292-297.

[63] Qureshi N, Blaschek HP. Economics of butanol fermentation using hyper-butanol producing Clostridium beijerinckii BA101. Food Bioprod Process 2000, 78(C3): 139-144.

[64] QureshiN, MeagherMM, HutkinsRW. Recovery of butanol from model solutions and fermentation broth using a silicalite silicone membrane. J MembranE Sci. 1999a, 158(1-2): 115-125.

[65] Qureshi N, Blaschek HP. Production of acetone butanol ethanol (ABE) by a hyper-producing mutant strain of Clostridium beijerinckii BA101 and recovery by pervaporation. BiotechnolProg 1999b, 15:594–602.

[66] Qureshi N, Saha BC, Dien B, Hector RE, Cotta MA. Production of butanol (a biofuel) from agricultural residues: part I—use of barley straw hydrolysate. Biomass Bioenerg. 2010a, 34:559–565

[67] Qureshi N, Blaschek HP. Recovery of butanol from fermentation broth by gas stripping. " Renewable Energy. 2001b, 22(4): 557-564.

[68] Qureshi N, Blaschek HP: Butanol production from agricultural biomass. In Food Biotechnology. Edited by Shetty K, Pometto A, Paliyath G. Boca Raton, FL: Taylor & Francis Group plc; 2005a, 525-551.

[69] Qureshi N, Saha BC, Hector RE, Hughes SR, Cotta MA. Butanol production from wheat straw by simultaneous saccharification and fermentation using Clostridium beijerinckii: Part I – Batch fermentation. Biomass Bioenerg. 2008a, 32, 168-175.

[70] Qureshi N, Ezeji TC, Ebener J, Dien BS, Cotta MA, Blaschek HP. Butanol production by Clostridium beijerinckii. Part I: use of acid and enzyme hydrolyzed corn fiber. Bioresourse Technol. 2008b, 99:5915–5922.

[71] Qureshi N, Hughes S, Maddox IS, Cotta MA. Energy-efficient recovery of butanol from model solutions and fermentation broth by adsorption. Bioprocess Biosyst Eng. 2005b, 27(4):215-222.

[72] QureshiN, Maddox IS, Friedl A. Application of continuous substrate feeding to the abe fermentation - relief of product inhibition using extraction, perstraction, stripping, and pervaporation. Biotechnol Progr. 1992, 8(5): 382-390.

[73] Qureshi N, Maddox IS. Continuous production of acetone-butanol-ethanol using immobilized cells of Clostridium acetobutylicum and integration with product removal by liquid-liquid extraction. J Ferment Bioeng. 1995, 80(2):185-189.

[74] Qureshi N, Saha BC, Hector RE, Cotta MA. Removal of fermentation inhibitors from alkaline peroxide pretreated and enzymatically hydrolyzed wheat straw: Production of butanol from hydrolysate using Clostridium beijerinckii in batch reactors. Biomass Bioenerg. 2008c, 32(12):1353-1358.

[75] Qureshi N, Saha BC, Hector RE, Dien B, Hughes S, Liu S, Iten L, Bowman MJ, Sarath G, Cotta MA. Production of butanol (a biofuel) from agricultural residues: part II - use of corn stover and switchgrasshydrolysates. Biomass Bioenerg. 2010b, 35:559–669

[76] Reijnders L. Conditions for the sustainability of biomass based fuel use. Energy Policy. 2006, 34:863–876.

[77] Ren C, Gu Y, Hu SY, Wu Y, Wang P, Yang YL, Yang C, Yang S, Jiang WH. Identification and inactivation of pleiotropic regulator CcpA to eliminate glucose repression of xylose utilization in Clostridium acetobutylicum. Metab Eng. 2010, 12(5): 446-454.

[78] Schwarz WH, Gapes JR, Zverlov VV, Antoni D, Erhard W, Slattery M. Personal communication and demonstration at the TU Muenchen (Campus Garching and Weihenstephan) in June 2006

[79] Shaheen R, Shirley M, Jones DT. Comparative fermentation studies of industrial strains belonging to four species of solvent-producing Clostridia. J Mol Microbiol Biotechnol. 2000, 2(1): 115-124.

[80] Shen CR, Liao JC. Metabolic engineering of Escherichia coli for 1-butanol and 1-propanol production via the keto-acid pathways. Metab Eng. 2008, 10(6): 312-320.

[81] Sillers R, Al-Hinai MA, Papoutsakis ET. Aldehyde-alcohol dehydrogenase and/or thiolase overexpression coupled with CoA transferase downregulation lead to higher alcohol titers and selectivity in Clostridium acetobutylicum fermentations. Biotechnol Bioeng. 2009, 102(1): 38-49.

[82] Sillers R, Chow A, Tracy B, Papoutsakis ET. Metabolic engineering of the non-sporulating, non-solventogenic Clostridium acetobutylicum strain M5 to produce butanol without acetone demonstrate the robustness of the acid-formation pathways and the importance of the electron balance. Metab Eng. 2008, 10(6): 321-332.

[83] Sivagnanam K, Raghavan VGS, Shah M, Hettich RL, Verberkmoes NC, Lefsrud MG. Comparative shotgun proteomic analysis of Clostridium acetobutylicum from butanol fermentation using glucose and xylose. " Proteome Sci 2011, 9:66

[84] Sivagnanam K, Raghavan VGS, Shah M, Hettich RL, Verberkmoes NC, Lefsrud MG. Shotgun proteomic monitoring of Clostridium acetobutylicum during stationary

phase of butanol fermentation using xylose and comparison with the exponential phase. " J Ind Microb Biotechnol. 2012, 39(6): 949-955.

[85] Soni BK, Das K, Ghose TK. Bioconversion of agro-wastes into acetone butanol. Biotechnology Letters. 1982, 4(1):19-22.

[86] Sun ZJ, Liu SJ. Production of n-butanol from concentrated sugar maple hemicellulosic hydrolysate by Clostridia acetobutylicum ATCC824. BIOMASS & BIOENERGY. 2012, 39(SI):39-47.

[87] Tomas CA, Welker NE, Papoutsakis ET. Overexpression of groESL in Clostridium acetobutylicum results in increased solvent production and tolerance, prolonged metabolism, and changes in the cell's transcriptional program. " Appl Environ Microbiol. 2003, 69(8): 4951-4965.

[88] Tracy BP, Jones SW, Fast AG, Indurthi DC, Papoutsakis ET. Clostridia: the importance of their exceptional substrate and metabolite diversity for biofuel and biorefinery applications. Curr Opin Biotechnol. 2012, 23(3): 364-381.

[89] Tran HTM, Cheirsilp B, Hodgson B, Umsakul K. Potectial use of Bacillus subtilis in a co-culture with Clostridium butylicum for acetone-butanol-ethanol production from cassava starch. Biochem Eng. 2010, 48:260–267

[90] TummalaSB, Welker NE, Papoutsakis ET. Design of antisense RNA constructs for downregulation of the acetone formation pathway of Clostridium acetobutylicum. J Bacteriol. 2003, 185(6): 1923-1934.

[91] Wang L, Chen HZ. Increased fermentability of enzymatically hydrolyzed steam-exploded corn stover for butanol production by removal of fermentation inhibitors. Process Biochem. 2011, 46(2):604-607.

[92] Welch RW, Rudolph FB, Papoutsakis E. Purification and characterization of the NADH-dependent butanol dehydrogenase from Clostridium acetobutylicum (ATCC 824). Arch Biochem Biophys. 1989, 273(2): 309-318.

[93] XiaoH, LiZ, Jiang Y, Yang Y, Jiang W, Gu Y, Yang S. Metabolic engineering of d-xylose pathway in Clostridium beijerinckii to optimize solvent production from xylose mother liquid. Metab Eng. 2012, DOI: 10. 1016/j. ymben. 2012. 05. 003

[94] Yan J, Lin T. Biofuels in Asia. Appl Energy. 2009, 86:1–10.

[95] Youngleson JS, Lin FP, Reid SJ, Woods DR. Structure and transcription of genes within the β-hbd-adh1 region of Clostridium acetobutylicum P262. FEMS Microb Lett. 1995, 125(2–3): 185-191.

[96] Zheng YN, Li LZ, Xian M, Ma YJ, Yang JM, Xu X, He DZ. Problems with the microbial production of butanol. J IndMicrobiolBiotechnol. 2009, 36:1127-1138.

[97] Zhou A, Thomson E. The development of biofuels in Asia. Appl Energy 2009, 86:11–20.

[98] Zverlov VV, Berezina O, Velikodvorskaya GA, Schwarz WH. Bacterial acetone and butanol production by industrial fermentation in the Soviet Union: use of hydrolyzed agricultural waste for biorefinery. ApplMicrobiolBiotechnol. 2006, 71:587–97.

Chapter 7

Generation of Biohydrogen by Anaerobic Fermentation of Organic Wastes in Colombia

Edilson León Moreno Cárdenas,

Deisy Juliana Cano Quintero and

Cortés Marín Elkin Alonso

Additional information is available at the end of the chapter

1. Introduction

1.1. The trouble of organics solids wastes

In the protection of environment, the adequate handling of solids wastes occupy a main place, the integral handling of wastes is a term applied to all activities associated with the wastes management in the society. The main aim is the administration of wastes associated with the environment and public health. The handling of solid wastes is one of the main environmental problems in the cities due to its generations increase simultaneously with the growth of the cities, its industrialization and the increase of population. In addition, the actual life style carries out a high demand of consumption of goods that generally are thrown out in a short time; this generates more production of wastes and therefor having to search for solutions to the final disposition.

A solution for the trouble of the urban solids wastes is the implementation of process of reusing and giving value to the different materials that form what is known as "garbage", with the purpose of obtaining products or sub products that can be to introduce into new economic cycles. The maximization of reusing and giving value to solids wastes carry out benefits as: less consumption of natural sources, reduction of energy consumption, less environmental pollution, better use of the location where the garbage is placed and economic benefits from recovered materials. So the changes of consumption patron and the sustainable production are essential for the reduction of wastes production.

Is very difficult to stop the production of solids wastes, the idea is consider the solids wastes as a source of material reusable, raw matter, organics nutrients, biofuels and energetics fuel. The set of process to recover and treatment the wastes are known as valorization of solids wastes. This production of wastes is due to origin, social context and production activities [1]. During the valorization and reusing of wastes, is necessary take account aspect as recollection and transport, with this is possible to obtain highs benefices by the transformation. Additionally is necessary to include applications of new concepts related to the financial services, decentralized management, community contribution and the options of transformation, valorization and incorporation to economic cycles [2].

1.2. Source of wastes

At the whole world, the solids wastes from different sources are generating negative environmental impact to the nature, the biodiversity and life in the planet. This is caused by the inappropriate disposition of wastes, the increase of population, the processes of industrial transformation, agroindustrial and life habits of people [3]. At the present time, one characteristic of the society is the increase unbridled of the production and accumulation of solids wastes, which are generated without a solution to its final disposition. In the most of cases, this produced an inappropriate final disposition, an increase in the environment deterioration (air, surface water and groundwater, soil, landscape), problems in the public health and personal security [2].

The characteristics of solids wastes changed in function of the main activity (industry, trade, tourism and others), the habits of the population, type of fed, consumption models, environment conditions and others. The solids wastes can be classified according to: the source (domestic activities, institutional, commercial, industry, farming, municipal services and construction); the constitution (recyclable material and non-recyclable) and grade of danger (commons and dangerous).

The present chapter shows the energetic potential of the solid organic wastes generated in Colombia and its capacity to produce biohydrogen by anaerobic fermentation; additionally is presented a research carried out at the Laboratory of Agricultural Mechanization of the National University of Colombia in Medellín between the years 2009 and 2012, which main aim was determinate the initial feasibility to generate biohydrogen from urban organics wastes and to establish some conditions to operate a bioreactor type batch.

2. Generation of solids wastes in Colombia

The quantity of wastes produced depend of factors as: the number of inhabitant in the city, urbanization rate, consumption habits, cultural practices to handle of wastes, the income, the application of technology and industrial development. According to the information reported by the "Superintendencia de Servicios Públicos Domiciliarios" by 2008, see [4], in Colombia were generated daily 25.079 tons of urban solids wastes, 10 million of tons/year, which 77% were domestics (19.310,8 ton); 15% Industrials (3.761,9) and 8% others (2.006,3 ton). In the country, the management of wastes is focused to the final disposal in landfill; only 2,4% is dedicated to recycle and valorization [1].

Disposing of wastes	tons/day	Participation (%)	Municipalities
Landfill	22.204	88,5	653
Open dumpsite	2.185	8,7	297
Treatment facility	615	2,4	98
Buried	75	0,3	19
Discharged into rivers		<0,1	10
Incineration		<0,1	11
Total	25.076	100	1.088

Source: [4].

Table 1. Disposing of wastes in Colombia

In the country, the solid wastes are mainly composed of organic material (65%), followed by the plastics (14%), paper and cardboard (5%), glass (4%), other components with minor participation.

Type of waste	Percentage %
Organic	65
Paper and cardboard	5
Plastic	14
Glass	4
Rubber	1
Metals	1
Textiles	3
Dangerous and pathogens	2
Others	5

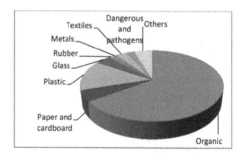

Source: [2].

Table 2. Composition of solids wastes in Colombia

In Colombia the major quantity of solid wastes generated are collected and treated by municipal companies (waste from domestic activities, commercials and industrials); however in some regions the problem of wastes solids is very important as the final disposition is made with little control, generating environmental pollution. The production of wastes (kg/habitant/day) is approximately 0,5 kg/habitant/day, oscillating between 1 kg/habitant/day for the big cities until 0,2 kg/habitant/day in the small towns [5]. The "Superintendencia de Servicios Públicos Domiciliarios" published by 2002 a study about the final disposition of the solids wastes in 1.086 cities. The technologies more frequent are: dumpsite and open incineration (52%), then landfill (30 %), and finally the use of composting, incineration and others (18%), [6].

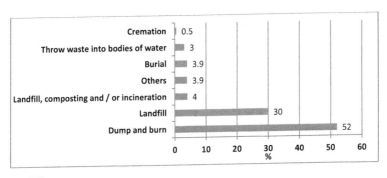

Source: [6].

Figure 1. Final disposition of solids wastes in Colombian for 1.086 municipalities, 2002.

There are two options to solve the problems generated by urban solid wastes which can be applied simultaneously to reach an optimum result:

- The first option according to the National Politics of Solid Wastes is give priority to integral management of solids wastes, focusing the operations management in the reuse and valorization of different materials that integrate the urban solid wastes.

- The second option is to take the wastes and give them an adequate final disposal in landfill operated technically.

The biomass in Colombia has calorific values between 4,384 kcal / kg for stems of coffee and 1,800 kcal / kg for banana rachis [7]. These values are comparable with reports from other countries as China where biomass from agricultural and forest activities have values between 3,827 to 4,784 kcal / kg [8]. In Argentina the lignocellulose biomass has values between 3,000 – 3,500 Kcal / kg and the municipal wastes between 2,000 and 2,500 Kcal / kg [9].

2.1. Colombian normativity about solid wastes

The Colombian normativity related to management of organics solids wastes began with the code of renewable natural sources (decree 2811 of 1974) and were implemented the followed norms:

- Decree 2104 of 1983: management of solids wastes.

- Resolution 2309 of 1986: special solids wastes.

- Law 142 of 1994: Law of public services.

- Decree 605 of 1996: Indications for an adequate cleaning service, from the generation, storage, collection, transport, to final disposition.

- Committee Technical ICONTEC 000019 about environmental management of solids wastes.

- Decree 1716 of August 2002 of "Ministerio de Desarrollo Económico" (In English: Economic Development Ministry) by mean the law 142 of 1994, law 632 of 2000 and the law 689 of 2001, related to the cleaning public service, the law 2811 of 1974 and the law 99 of 1993. The article 8 related to the program for the integral management of solids wastes, which should be realized by the cities in a maximum time of 2 years [10].

2.2. Organic solids wastes and its energetic potential in Colombia

The increase of energy demand in recent decades driven in particular by developed countries and countries with economic growth as Colombia, is leading to rapid depletion of nonrenewable energy resources, increasing pollution and global warming. The alternatives energetics sources emerge as a great option to reduce the adverse effects of this development. The biomass is considered as the alternative energetic source the most potential, according to reports from the World Energy Council [11] it is estimated that energy from biomass will account for 25,4% of global consumption by 2030 and 80% by 2080. Biomass is very varied due to its production and origin, a particular type are the wastes of natural processes, industrial or agroindustrial. It is estimated that Colombia has an energy potential from residual biomass of 449.485 TJ / year, also has a land area of 114,174,800 hectares, of which 44,77% are engaged to agricultural activities, this places to sector as the main source of wastes (with an energy potential of 331.645 TJ / year, mainly from annual and permanent crops). At the second place are the wastes from livestock activities (with an energy potential of 117.747 TJ / year), then the urban organic wastes (wastes from food and homes with an energy potential of 91 TJ / year) and finally the wastes from agroindustrial activities [12].

Among the methods to profit energetically the residual biomass, the anaerobic fermentation is a way of great interest, with this bioprocess is possible to generate a gas with high energy characteristics such as hydrogen and sludge that could be employed as fertilizer on crop. The generation of biohydrogen by anaerobic fermentation of wastes has generated great interest in the last decades. Hydrogen is a promising option as energy source [13, 14], it is a clean renewable resource because its combustion produces only water as emissions, in addition has the highest energy content per unit mass, with a value of 122 kJ / g [13]. The biological production of hydrogen can be seen as a promising option [15], two types of bacteria are involved in the process: acidogenic bacteria which initially to reduce the substrate in H_2 (biohydrogen), acetic acid and CO_2 and the methanogenic bacteria that converted these elements in methane gas. If the purpose is to produce biohydrogen, favorable conditions for the growth of the first type of bacteria (acidogenic) should be provided, inhibiting or eliminating the population of methanogenic bacteria [16]. Currently there are two methods to inhibit this type of bacteria: thermal shock and acidification [17, 18].

The residual biomass in Colombia has a high potential as alternative energetic source, only in wastes of sugar cane, rice husk, coco fiber, coffee pulp, oil palm, bean seed and barley, the potential is 12.000 MW/year approximately. The wastes are produced in different regions of the country and during all year. The country has a potential for generation of biomass of 331'638.720 ton/year, if all agricultural and urban wastes were treated by fermentation anaerobic, could be generated 28'825.609 m³ of biohydrogen, this might give a energetic potential of 144 GW, upper value to country potential in wind energy (21 GW), tidal energetic potential (30 GW with two coasts) and geothermic energetic potential (1 GW).

In Colombia this quantity of biohydrogen could replace all diesel requested by the diesel electrical plants installed in the country. This has a great important especially in regions without connection to national electrical grid. In the country approximately the 66% of the territory are not connection to national electrical grid, this is 1,4 millions of people, namely the 4% of the population. The country has an installed electric capacity at the region without connection to national electrical grid of 102 MW of which 97 MW are produced by diesel plants, this quantity could be generated, using only the 40% of the urban organic wastes generated at the country. Colombia produces 250.000 tons/year of banana wastes with a potential to generated 100.000 m³ of biohydrogen by anaerobic fermentation, this represent 500 MW of energy per year, quantity enough to supply the electric energy demand of 200.000 people during a year.

3. Generation of biohydrogen in Colombia

A research in order to determine the initial feasibility to generate biohydrogen from urban organics wastes and then established some conditions to operate a batch bioreactor was developed in Colombia. This section presents the results of this research and analysis the potential use of urban wastes as sources to generate hydrogen.

3.1. Localization

The research was performance between the years 2009 and 2012, at the Laboratory of Agricultural Mechanization of the National University of Colombia in Medellín, localized in 6°13'55"N and 75°34'05"W, with average annual temperature of 24°C, relative humidity of 88% and average annual precipitation of 1571mm.

3.2. Methods

Two stages were established to develop the research, the first had five phases.

3.2.1. First stage

Phase 1. Identification of organic wastes generated at the Central Wholesaler of Antioquia

The Central Wholesaler of Antioquia is the main company dedicated to trade food in the city of Medellín (fruits, vegetable and some grains). At the first phase historical information related to organic wastes production during two year was supplied by Central Wholesaler of Antioquia and was made a photographic register of solids wastes generated. The photographs were taken twice per day at the morning and afternoon.

Phase 2. Selection of wastes with greater production

According to the information collected and the photographic register from the first phase, the wastes with greater production were selected to be introduced into a batch bioreactor.

Phase 3. Elemental Composition and chemical composition analysis

The quantity of volatile solids, total solids and elemental composition on both wet and dry basis (coal, nitrogen and hydrogen) were obtained for each wastes. Were taken samples of 5 grams and the analysis method applied was the Wendee method (the analysis was made at the chemical analysis laboratory of National University in Medellin). With that information was calculated the quantity of wastes to use. Six samples of 3 grams in each wastes were taken in order to obtain the elemental analysis, in this case the method applied was burn of sample and the equipment employed was an elemental analyzer CE – 440 (Figure 2a). The samples were triturated with a precision crusher – IKA WERNE with sieve of 0,5 mm (Figure 2b) and then were dried in a lyophilizer LABCONCO Freezone 12L (Figure 2c). In order to determine the quantity of wastes and water to be employed, 6 grams of volatile solids per liter-day were used as organics load [19], additionally was employed a concentration on volatile solids of 5% [20].

Figure 2. Elemental analyzer CE – 440 (a), Crusher MF Basic- IKA WERKE (b), Lyophilizer LABCONCO - Freezone 12L (c).

Phase 4. Installation of bioreactor

A batch bioreactor of 2000 liters was installed, the wastes were triturated to facilitate its access into bioreactor and its process by the bacteria. The quantity of gas generated was registered with a gas flow meter Metrex G 2,5 with accurate of 0,040 m^3/h; maximum pressure of 40 kPa, additionally was employed a gel of silica to remove the wet of gas. The load of bioreactor was made during four days, each day was used the same quantity until to complete the total load.

Phase 5. Principal variables to register

The relativity humidity and environment temperature were registered daily, was used a thermohygrometer with rank in temperature until 120°C and 100% in relativity humidity (Figure 4). The pH into the bioreactor was registered daily too, in this case was employed a digital pH-meter Hanna Instruments, with accurate of ± 0,2 (reference temperature of 20°C).

Figure 3. Installation of bioreactor and equipment to trituration

Figure 4. Thermohygrometer and pH-meter

The organics load was determined at the beginning and end of bioprocess; in this case the total suspended solids (TSS), total solids (TS), volatile fatty acids (VFAs), chemical oxygen demand (COD) and biochemical oxygen demand (BOD) were determined. The analytics method employed were Standard Method by water and residual water of the APHA-AWWA-WPCF, edition 19 of 1995.

The production of gas was registered daily, samples were collected in Tedlar bags (with capacity of 1 liter, Figure 5) and then were analyzed in a chromatographic gas (Perkin Elmer) to establish its composition (percentage of CO_2, O_2, H_2, CH_4 and N_2). During the tests, the wastes were subjected to an acid pretreatment to eliminate the methanogenic bacteria, after several days, agricultural lime was added to increase the pH until to obtain a value most adequate to the acidogenic bacteria.

3.2.2. Second stage

With the information from the first stage was elaborated an experiment with three treatments and three repetitions, were used three bioreactors of 2000 liters each them. The treatments were integrated by three values of duration to acid pretreatment (3, 7 and 10 days) and three values of pH to operation of the bioprocess (4,5 – 5,0; 5,1 – 5,5 and 5,6 – 6,0). The materials used were wastes of different fruits and vegetal from Central Wholesaler of Antioquia. The wastes were

Figure 5. Gas flow meter and Tedlar bag

triturated and mixed with water in a relation of 1:2,5. In each test were taken samples of wastes and sludge to determinate the organic load, in addition was recorded daily the pH and the gas production. When the pretreatment of acidification ended, agricultural lime was added like at the first stage. The methodology employed to obtain the quantity of gas generated was the same of the first stage, was used a gas flow meter Metrex G2,5 with accurate of 0,040 m³/h and maximum pressure of 40 kPa. Samples of gas were collected in Tedlar bags and then were analyzed in a chromatographic gas (Perkin Elmer) to determinate its composition.

The organics load of wastes was obtained at the beginning and end of bioprocess; this included the total suspended solids (TSS), total solids (TS), volatile fatty acids (VFAs), chemical oxygen demand (COD) and biochemical oxygen demand (BOD). The analytic method employed was the Standard Method by water and residual water like the first stage. Was calculated the production of gas (liters/day), hydrogen percentage (% de H_2) and yield of biohydrogen (liters of H_2/day).

3.3. Results

3.3.1. First stage

First and second phase: The quantity and percentage of wastes generated at the Central Wholesaler of Antioquia during the year 2011 are show at the Table 3 and Figure 6. The highest production of wastes was associated to cabbage and lettuce leaves then wastes of citrics (orange and lemon) and finally wastes of mango, guava and others tropical fruits. The Figure 7 shows some pictures of wastes in the storage containers at the Central Wholesaler of Antioquia. With the information of production were selected wastes of cabbage and lettuce leaves, orange, mango, papaya and guava to be employed at the bioprocess.

Third phase: The elemental analysis of wastes selected is show at the Table 4. To orange wastes, the relation C/N obtained was less than values reported in others research. In the others cases the results were close to values reported for wastes with similar characteristics. A relation C/N close to 30 is considered appropriate to growth of anaerobic bacteria [22].

Organic wastes	Volumen average month (m³)
Lettuce and cabbage leaves	360
Orange and lemon	38
Pimento, cucumber	26
Mango	23
Tomato	23
Papaya and guava	22
Total	492

Source: [21].

Table 3. Organics wastes generated at the Central Wholesaler of Antioquia

Organics waste

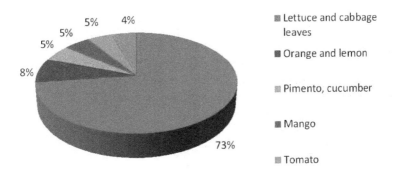

Figure 6. Percentage of wastes generated at the Central Wholesaler of Antioquia, year 2011

Wastes	C	H	N	C/N	C/N (Literature)
Mango	37.6	6.0	1.5	25.9	34.8
Orange	40.6	5.5	1.3	31.5	75.6
Guava	40.9	5.7	1.4	29.2	34.8
Papaya	36.7	5.7	1.3	27.6	34.8
Lettuce and cabbage leaves	37.6	5.3	1.5	25.1	18.0

Table 4. Result of elemental analysis on dry basis (%), (Coil laboratory, National University of Colombia)

Figure 7. Pictures of wastes in the storage containers at the Central Wholesaler of Antioquia

The chemical composition analysis of wastes showed that the highest values of volatile solids were found in the tropical fruits (mango, orange, guava and papaya). The volatile solids are the proportion of the raw material that bacteria using to generate biogas and have an outstanding role during the anaerobic fermentation process.

Waste	ST (%)	SV (%ST)	SV (%)
Mango	97,4	15,1	14,71
Orange	96,6	14,3	13,81
Guava	96,7	15,3	14,79
Papaya	97,1	12,7	12,33
Lettuce and cabbage leaves	86,5	8	6,92

Table 5. Chemical composition analysis, (Chemical composition analysis laboratory, National University of Colombia)

In order to determinate the quantity of wastes to be used was obtained the density of each wastes, to this were taken samples and then were triturated, weighed and finally was calculated the volume to employ. The bioreactor was loaded with 422 kilograms of wastes and 1110 kilograms of water, this provided an average relation (wastes: water) of 1:2,5.

Wastes	Density (kg/l)	So (g SV)/l	Organic load (g SV)/day [19]	Wastes to use (l)	Wastes to use (kg)	Concentration of volatile solid (% SV)/day [20]	Relation (wastes: water)	Water to use (l)
Mango	0,8820	129,7	6,0 .	22	19	5	1 : 2.94	64
Orange	0,9639	133,1	6,0	22	21	5	1 : 2.76	61
Guava	1,1655	172,4	6,0	29	33	5	1 : 2.96	85
Papaya	1,1907	146,8	6,0	24	29	5	1 : 2.47	60
Lettuce and cabbage leaves	0,4579	31,7	6,0	5	2	5	1 : 1.38	7
Total				102	105			278
Total to 4 days				409	422			1110

Table 6. Quantity of wastes and water to the fermentation process

Fourth phase:

Each waste was triturated and mixed with water during three minutes until to reach an average size of 2 centimeters. In order to reduce the quantity of methanogenic bacteria, the wastes were submitted to acidic conditions during three months with a value of pH close to 3,5. Afterwards was added during three days agricultural lime until to reach a pH of 6,2; in that moment the production of biohydrogen started. The quantity of agricultural lime added was 7 kilograms (Figure 9).

Figure 8. Bioreactor used by the first stage and wastes triturated

Fifth phase:

The organics load showed an important reduction during the process, the total suspend solids were reduced in 83%, the chemical oxygen demand was reduced in 65% and the biochemical

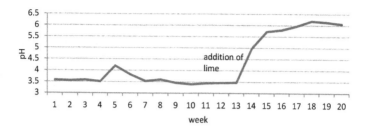

Figure 9. Behavior of pH during the first stage

oxygen demand was reduced in 63,6%. The environment temperature was between 21,8 y 31 °C, this mean that the biohydrogen production was developed under mesophilic conditions. The average relative humidity was between 38 y 73%.

Analysis	Beginning	End
SST (mg/l)	1920	325
STV(mg/l)	54815	8296
ST(mg/l)	62395	9893
COD(mg/lO$_2$)	54000	19133
BOD(mg/lO$_2$)	37633	13713

Table 7. Organic load of wastes at the first stage (Laboratory of Sanitary Engineering, National University of Colombia)

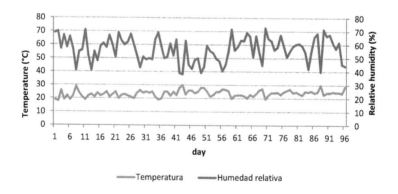

Figure 10. Behavior of temperature and relative humidity average

The gas production started three days after application of agricultural lime and continued for 22 days more. The hydrogen (biohydrogen) percentage found in gas ranged between 6,37 y 17,26; with a percentage of hydrogen less than 13,3; there was carbon dioxide and nitrogen in the biogas, however when the percentage of hydrogen was greater than 13,3; the gas composition was only hydrogen and carbon dioxide. The greater value of methane was 1,25% and less was 0%, this mean that the pretreatment to reduce the methanogenic bacteria was satisfactory.

Sample	CO_2 (%)	H_2 (%)	N_2 (%)	O_2 (%)	CH_4 (%)
1	31,79	6,72	48,19	13,06	0
2	70,99	13,31	2,63	0,42	1,25
3	75,67	17,26	0,65	0,096	0,73
4	80,98	13,51	ND	ND	0,6
5	32,80	6,37	48,13	13,16	0,24

ND: not detected

Table 8. Composition of gas generated (Coil laboratory, National University of Colombia)

The total production of hydrogen was 177 liters in 22 days, with a maximum value of 14,5 liters, an average of 7,4 liters of H_2/day and maximum yield of 83 liters of H_2/m³ of bioreactor. The maximum value of generation of hydrogen was registered 7 days after from started the gas production and the maximum rate of hydrogen generation was obtained between first and seventh days. The Figure 12 shows a several pictures of biohydrogen generated, the color blue is from silica gel used to remove the wet of the gas. The quantity of organic load removed was 26.400 mg/liter of O_2, (COD).

Source: Information personal from research

Figure 11. Pictures of biohydrogen generated

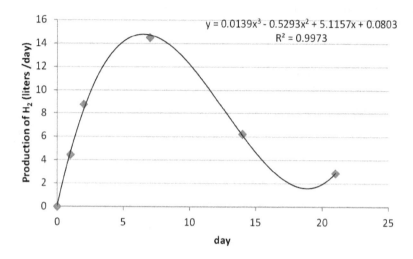

Figure 12. Production daily of hydrogen

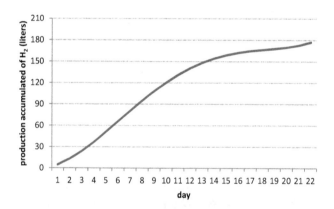

Figure 13. Production accumulated of hydrogen

3.3.2. Second stage

Installation of bioreactors and wastes to use

At the second stage were used the same wastes of first stage, but additionally were employed wastes of tomato, onion, garlic and husk of cape gooseberry (Table 10). The wastes were triturated and mixed with water during three minutes, the relation of wastes: water was 1:2,5 like at first stage. The quantity of wastes employed in each treatment was similar to quantity used at first stage. The volume of work in each bioreactor was 70% and 30% was dedicated to storage the gas generated.

Figure 14. Set of bioreactors employed at the second stage

Repetition	Wastes	Quantity of wastes (kg)		
		T1	T2	T3
1	Lettuce and cabbage leaves, tomato, onion, garlic, pimento, orange, lemon, mango, guava and papaya	506	514	453
2	Lettuce and cabbage leaves, tomato, onion, garlic and husk of cape gooseberry	450	450	450
3	Lettuce and cabbage leaves, orange, mango, guava and papaya	500	500	400
Average		485,3	488	434,3

Table 9. Wastes used in each repetition

The highest values of chemical oxygen demand (COD) were obtained in the treatment 2 during the repetition 1 and the treatments 1 and 3 of the repetition 3 respectively, namely in these cases there were more quantity of food available to the microorganisms.

Analysis	Repetition 1			Repetition 2	Repetition 3		
	T1	T2	T3	T1, T2 y T3	T1	T2	T3
ST (mg/l)	5322	8198	4700	11280	16290	9830	10500
COD (mg/ IO_2)	8667	18000	8000	12000	27140	17940	23340
BOD (mg/IO_2)	7617	10983	6200	5840	24415	3775	14455

Table 10. Organic composition of wastes employed

Behavior of pH during the pretreatment and operation of bioreactors

Due to type of wastes employed in the repetition 2, was necessary to add muriatic acid into all bioreactors to achieve the pH of acidification, but there were not response (pH between 3,5 and 4,5). However, during the repetitions 1 and 3, in all treatments was used wastes of orange and lemon, this allowed to apply the pretreatment of acidification, afterwards was added agricultural lime and was reached a pH between 5 and 6, values adequate to generate biohydrogen.

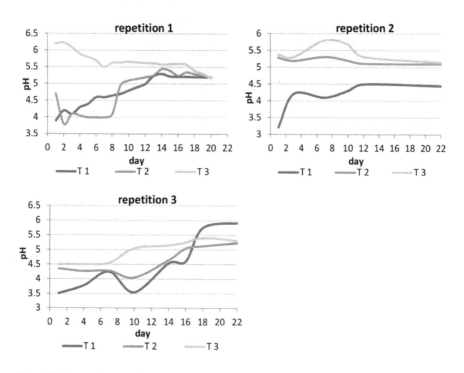

Figure 15. Behavior of pH in each treatment and repetition

Production of biohydrogen

The gas generated in all treatments was compound of hydrogen, carbon dioxide, nitrogen and oxygen. The greater value of methane was 3,7% and the less was 0%, in many times the methane was not detected (ND), this mean that the pretreatment to reduce the methanogenic bacteria was satisfactory. The percentage of hydrogen in gas was between 5 and 18,08; this was the highest value in the research and was obtained in the treatment 3 during the repetition 3 when the wastes used were Lettuce and cabbage leaves, orange, lemon and papaya. In the repetition 2 in all treatments, there were not generation of hydrogen (NG). The oxygen content in some repetitions show maybe that some air entered to bioreactor when the samples were taken.

Treatment	Sample	CO_2 (%)	H_2 (%)	N_2 (%)	O_2 (%)	CH_4 (%)
	1	35,1	6,7	53,9	3,0	1,4
1	2	16,2	6,4	64,8	13,6	ND
	3	18,2	0,02	67,2	13,9	0,8
	1	40,0	10,5	45,7	0,9	0,8
2	2	46,5	7,5	43,7	1,8	ND
	3	34,1	0,02	54,3	7,0	3,7
	1	40,2	7,7	45,0	2,5	0,4
3	2	43,9	7,0	47,0	2,1	ND
	3	4,0	0,02	77,3	18,1	ND

Table 11. Composition of gas generated, first repetition

Treatment	Sample	CO_2 (%)	H_2 (%)	N_2 (%)	O_2 (%)	CH_4 (%)
	1	NG	NG	NG	NG	NG
1	2	NG	NG	NG	NG	NG
	3	NG	NG	NG	NG	NG
	1	11,4	5,5	41	13,4	0,2
2	2	37,8	5,5	36,2	5,4	0,5
	3	20,1	5,7	43,5	12	3,3
	1	NG	NG	NG	NG	NG
3	2	NG	NG	NG	NG	NG
	3	NG	NG	NG	NG	NG

Table 12. Composition of gas generated, second repetition

Treatment	Sample	CO_2 (%)	H_2 (%)	N_2 (%)	O_2 (%)	CH_4 (%)
	1	29,08	7,5	50,3	3,0	0,2
1	2	23,88	6,5	41,7	10,3	2,6
	3	42,14	8,2	36,1	7,1	1,4
	1	28	7,0	52,2	4,4	0,1
2	2	45,29	5,8	38,1	3,0	0,1
	3	40,69	5,3	40,8	3,2	0,1
	1	8,96	18,0	51,3	12,1	0,2
3	2	43,32	6,3	36,5	7,1	0,7
	3	52,85	5,0	30,5	5,1	0,3

ND: not detected
NG: not generated

Table 13. Composition of gas generated, third repetition

The maximum production of biohydrogen per day obtained at the first stage was 15 liters, however at the second stage the maximum production was 38 liters, this mean that the production was duplicated during the second stage. In the repetition 3, in the treatment 3 were generated 32 liters of biohydrogen, meanwhile in the treatment 1 were generated 15 liters. The wastes employed in both treatments were Lettuce and cabbage leaves, orange, lemon and tropical fruits as mango, guava and papaya, the initial pH was lesser to 4,5 during 7 and 8 days, and the pH of bioreactors operation was between 5 and 5,5.

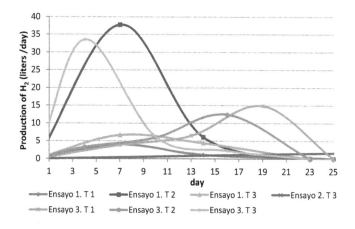

Figure 16. Production of hydrogen

The greater accumulated production was reached in the treatment 2 and the repetition 1, followed by the treatment 3 in the repetition 3. In both cases were used vegetal wastes and tropical fruits in same proportions in addition, initially the wastes were subjected to acid conditions with a pH less to 4,0 during 8 days and a pH for operation of bioreactor between 5 and 5,5. Under those conditions the hydrogen content into gas ranged between 7,5 and 10,5%. The total production of hydrogen in the treatment 2 and the repetition 1 was 317,8 liters in 22 days, with 14,44 liters of H_2/day (twice the result from the first stage), and maximum yield of 159 liters of H_2/m^3 of bioreactor. Other outstanding result was reached when were employed the same wastes, at beginning were applied acid conditions during 7 days under a pH less to 4,5; and then was used a pH for the operation of bioreactor between 5 and 5,5. In this case the percentage of hydrogen into gas ranged between 5 and 18,08% (the last value was the maximum content reached in the research). The production of hydrogen was 231,1 liters of H_2 in 22 days with 10,5 liters of H_2/day (duplicated the value reached from the first stage) and a maximum yield of 115,6 liters of H_2/m^3 of bioreactor.

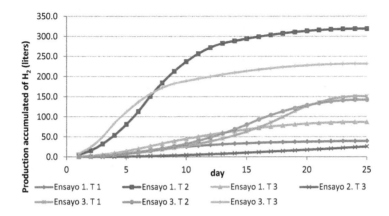

Figure 17. Production accumulated of hydrogen

4. Final analysis

Two treatments showed the highest production of biohydrogen, the treatment 2 in the repetition 1 and the treatment 3 in the repetition 3, the maximum value was obtained with the treatment 2 in the repetition 1 in which were used wastes from lettuce and cabbage leaves, tomato, onion, garlic, pimento, orange, lemon, mango, guava and papaya. The acid conditions were implemented 8 days with value of pH near to 4, the operation of bioreactor was between 5 and 5,5. In the treatment 3 in the repetition 3 were used the same wastes, the acid condition was applied during 7 days with value of pH near to 4,5; the pH of bioreactor operation was between 5 and 5,5. Although was generated more quantity of hydrogen in the treatment 2 during the repetition 1, was in the treatment 3 in the repetition 3 where was obtained the greater hydrogen content in the gas (18,04%) and greater rate of generation of hydrogen.

The maximum production of hydrogen was obtained at the second stage when the pretreatment of acidification was applied during 8 days with a value for the pH of 4, a pH of reactor operation between 5 and 5,5; and a value of chemical oxygen demand (COD) near to 20.000 mg/liter of O_2. At the first stage when was used a quantity of wastes from tropical fruits greater than wastes of lettuce and cabbage leaves, the chemical oxygen demand (COD) initial was 54.000 mg/liter of O_2, however the hydrogen production was significantly less respect to second stage. This indicates that a high value of chemical oxygen demand could inhibit the hydrogen generation; this result is according to reports of different authors [13, 23-29]. When were used vegetal wastes (without wastes of tropical fruits) as lettuce and cabbage leaves, tomato, onion, garlic and husk of cape gooseberry, there were no acid conditions at beginning the process and was necessary to add acid, however there was no response in the biosystems and the pH always was upper than 4,5. Under these conditions there was no production of hydrogen. In addition the chemical oxygen demand was low (12.000 mg/liter of O_2).

The results shows that is feasible to produce biohydrogen (hydrogen) when are employed organic wastes from the Central Wholesaler of Antioquia. The wastes should be submitted to a pretreatment acid with a pH between 3,5 y 4,0; during 7 days (or less), then the operation pH should be increased until a value between 5 and 5,5. The chemical oxygen demand (COD) should be between 20.000 and 54.000 mg/liter of O_2, this is possible to reach when in the bioprocess is employed a proportion similar of tropical fruit waste and vegetal waste.

5. Conclusions

- It was possible to generate hydrogen from organic wastes of Central Wholesaler of Antioquia and to improve the bioprocess.

- The chemical oxygen demand (COD) promoted the biohydrogen production, the best results were obtained to values between 20.000 and 54.000 mg/liter of O_2. These values were achieved with a heterogeneous mix of fruits and vegetal wastes.

- There was not generation of biohydrogen when the bioprocess started with a pH upper than 4. This ratifies that to generate biohydrogen by anaerobic fermentation is necessary to apply a pretreatment, in this research, a pretreatment under acid conditions (pH between 3,5 and 4,0) was successful.

- Colombia has a high potential to generate hydrogen by anaerobic fermentation due to organic wastes available, these wastes could generate until 28'825.609 m^3 of biohydrogen and supply an energetic potential of 144 GW, value upper than the installed potential (13,5 GW).

- The results show that is possible to produce biohydrogen by anaerobic fermentation of organic wastes and providing new sources energetic.

Acknowledgements

The authors acknowledge to National University of Colombia in Medellín and the Central Wholesaler of Antioquia by the financial support to research.

Author details

Edilson León Moreno Cárdenas*, Deisy Juliana Cano Quintero and
Cortés Marín Elkin Alonso

*Address all correspondence to: elmorenoc@unal.edu.co

Engineering Agricultural Department, National University of Colombia, Medellín, Colombia

References

[1] Contreras C. Manejo integral de aspectos ambientales – residuos sólidos. 2006. Diplomado gestión ambiental empresarial para funcionarios de ETB. Universidad Pontificia Javeriana, Bogotá, Colombia.

[2] Ministerio de Ambiente, Vivienda y Desarrollo Territorial. Viceministerio de Ambiente. Construcción de criterios técnicos para el aprovechamiento y valorización de residuos sólidos orgánicos con alta tasa de biodegradación, plásticos, vidrio, papel y cartón. Manual 1: generalidades. Bogotá: epam s.a. e.s.p; 2008.

[3] Acurio G, Rossin A, Teixeira F.P, and Zepeda F. Diagnóstico de la situación de residuos sólidos municipales en América Latina y el Caribe. Banco Interamericano de Desarrollo y la Organización Panamericana; 1997. http://www. Bvsde.paho.org/acrobat/diagnost.pdf (accessed 12 June 2012).

[4] Superintendencia de Servicio Públicos Domiciliarios. Estadísticas 2008. Bogotá; 2008. http://www.superservicios.gov.co. (accessed 21 April 2012).

[5] Puerta E. S. M. Los residuos sólidos municipales como acondicionadores de suelos. Revista lasallista de investigación. 2004; 1(1) 56 - 65.

[6] Superintendencia de Servicio Públicos Domiciliarios. Estadísticas 2002. Bogotá; 2002. http://www.superservicios.gov.co. (accessed 09 July 2012).

[7] Ministerio de Minas y Energía. Estadísticas: MME. http://www.minminas.gov.co/minminas (accessed 14 May 2012).

[8] Cuiping L, Chuangzhi W, Yanyongjie, Haitao H. Chemical Elemental Characteristics of Biomass Fuels in China. Biomasss and Bioenergy 2004; 27 (2) 119-130.

[9] Secretaria Argentina de Energía: Biocombustibles. http://energia3.mecon.goc.ar/home (accessed 13 August 2012).

[10] Ministerio de Desarrollo Económico. MCIT. http://www.mincomercio.gov.co/ (accessed 10 March 2012).

[11] World Energy Council. WEC. http://www.worldenergy.org/activities/knowledge_networks/energy_efficiency/default.asp (accessed 12 June 2012).

[12] Unidad de Planeación Minero Energética. UPME. http://www1.upme.gov.co (accessed 27 March 2012).

[13] Hallenbeck C. P. Fermentative Hydrogen Production: Principles, progress and prognosis. International Journal of Hydrogen Energy 2009; 34(17) 7379-7389.

[14] Lee Z. K, Shiue L. L, Jian S. L, Yu H. W. Effect of pH in fermentation of vegetable kitchen wastes on hydrogen production under a thermophilic condition. International Journal of Hydrogen Energy 2008; 33(19) 5234-5241.

[15] Vázquez D. G, Navarro C. C. B, Colunga R. L. M, Rodríguez A. L, Flores R. E. Contin-
      uous biohydrogen production using cheese whey: Improving the hydrogen production
      rate. International Journal of Hydrogen Energy 2009; 34(10) 4296-4304.

[16] Lee Y.W, Chung J. Bioproduction of hydrogen from food wastes by pilot-scale com-
      bined hydrogen/methane fermentation. International Journal of Hydrogen Energy
      2010; 35(21) 11746-11755.

[17] Cano Q.D.Y, Moreno C.E.L. Viabilidad del Aprovechamiento de Coproductos como
      fuente de combustible para una Celda de hidrógeno en la central mayorista de
      Antioquia. Tesis de Grado en Ingeniería Agrícola. Universidad Nacional de Colombia
      Sede Medellín, 2010.

[18] Ozmihci S, Kargi F. Dark fermentative bio-hydrogen production from wastes wheat
      starch using co-culture with periodic feeding: Effects of substrate loading rate. Inter-
      national Journal of Hydrogen Energy 2011; 36(12) 7089-7093.

[19] Shin H. S, Youn J. H, Kim S. H. Hydrogen production from food waste in anaerobic
      mesophilic and thermophilic acidogenesis. International Journal of Hydrogen Energy
      2004; 29(13) 1355-1363.

[20] Kim D.G, Kim S.H, Shin H.S. Hydrogen fermentation of food wastes without inoculum
      addition. Enzyme and Microbial Technology 2009; 45(3)181-187.

[21] Central Wholesaler of Antioquia. Wastes production, statistics 2011. Document interno.
      20011.

[22] Piedrahita V. D. R. Elementos para una tecnología sobre la producción de biogas.
      Medellín: Universidad Nacional de Colombia; 2000.

[23] Bouallagui H. O, Haouari Y, Touhami R, Ben C, Marauani L, Hamdi M. Effect of
      Temperature on the Performance of an Anaerobic Tubular Reactor Treating Fruit and
      Vegetal Waste. Process Biochemistry 2004; 39 (12) 2143-2178.

[24] Chenlin L, Fang H. Fermentative hydrogen production from wastewater and solid
      wastes by mixed cultures. Critical Reviews in Environmental Science and Technology
      2007; 37(1) 1-39.

[25] Das D, Veziroglu T.N. Advances in biological hydrogen production processes. Inter-
      national Journal of Hydrogen Energy 2008; 33(21) 6046-6057.

[26] Das D, Veziroglu T.N. Hydrogen production by biological processes: a survey of
      literature. International Journal of Hydrogen Energy 2001; 26(16) 13-28.

[27] Hawkes F.R, Hussy I, Kyazze G, Dinsdale R, Hawkes D.L. Continuous dark fermen-
      tative hydrogen production by mesophilic microflora: Principles and progress.
      International Journal of Hydrogen Energy 2007; 32(2) 172-184.

[28] Hwanga J.J, Chang W.R. Life-cycle analysis of greenhouse gas emission and energy
      efficiency of hydrogen fuel cell scooters. International Journal of Hydrogen Energy
      2010; 35(21) 11947-11956.

[29] Nishio N, Nakashimada Y. High rate production of hydrogen/methane from various substrates and wastes. Advances in Biochemical Engineering/Biotechnology 2004; 90 63-87.

Chapter 8

Coproducts of Biofuel Industries in Value-Added Biomaterials Uses: A Move Towards a Sustainable Bioeconomy

S. Vivekanandhan, N. Zarrinbakhsh, M. Misra and A. K. Mohanty

Additional information is available at the end of the chapter

1. Introduction

World population is expected to grow nearly 9 billion in 2040 and eventually increases the global energy demand by 30% compared to current conception [1]. The issues related to increasing trend of crude oil cost, depleting source of fossil fuels and emerging threat on greenhouse gas emissions are leading the global energy sector to undergo a fundamental transformation towards renewable energy sources [1-2]. As the result, a main focus is motivated on renewable energy technologies that are based on solar, wind and biofuels. In transportation point of view, biofuels receive extensive attention due to their versatility in storage and refilling. Both bioethanol and biodiesel come together as biofuel currently produced from renewable resources through two different pathways. In some countries like Brazil, biofuels are produced and marketed at competitive cost compared to petroleum-based fuels employing existing technology [3-4]. They also carry following advantages comparing to petro fuels; (i) create significantly less pollutants (SO_x and NO_x), which also mitigates CO_2 emission, (ii) biodegradable nature lead to the less environmental leak risk and (iii) provides better lubricant effect, which enhances the engine life [5]. In addition, these emerging biofuel technologies will be expected to create more economic benefits to agriculture sectors and new rural job opportunities. Moreover, biofuels are attractive options for future energy demand since they can be produced domestically by many countries while the respective retail and consumer infrastructure needs minimum modification; so does the existing engine and fueling technology [6].

However, biofuel foresees a challenging journey to benefit from its highest potentials and to guarantee a viable future. Primarily, it needs policy support and commercialization. At the same time, research and development is crucial to conquer the challenges and bring sustain-ability to biorefinery facilities [7]. Major motivation for biofuels usage arises from the execution of biofuel policies by many countries, which mandates the incorporation of bio-counterpart into traditional fuels. United Kingdom introduced the Renewable Transport Fuels Obligation (RTFO) and encouraged the oil suppliers to incorporate biofuel into transport fuel between 2.5 and 5% during 2008-2010. RTFO's ultimate aim is to increase this 5% upto 10% by 2020, which will reflect in the demand of minimum 5 million tonnes [8]. Renewable fuel blending mandates in Canada was implemented through Canadian Environmental Protection Act, which recom-mends 5% ethanol with gasoline (in 2010) and 2% biodiesel with diesel (in 2012) [9]. In South Africa, the National Biofuels Industrial Strategy was introduced by the government in 2007, which recommends the implementation of 2% biofuels into liquid road transport fuels by 2013 [10]. Currently India's ~80% crude oil demand is satisfied by foreign suppliers, which is projected to rise 90% in 2025. In order to reduce this foreign dependency, India has announced the target of ethanol blending with gasoline 20% by 2017 [11]. In biofuel production, China has clear production goals to meet emerging demand in near future. China's integrated biofuel polices (rural welfare, improved energy security, reduced fossil fuel dependence, and CO_2 emissions) aimed to meet ~ 15% of the total transportation fuel demand by 2020 [12]. In Malaysia, the National Biofuel Policy initially planned to proceed with 5% biodiesel blend with 95% petroleum diesel, which is similar to Europe's B5, which has been started from 2009. This will be implemented through short, medium and long term strategies aiming to reduce their petroleum imports [13]. In addition to that, many countries have already designed various incentive programs for the effective promotion of biofuel production including bioethanol and biodiesel. This implements 5-20% biofuel supplement into traditional fuels [14].

Such blending mandates of biofuels adopted by the E.U. and U.S. created a dispute of increased food prices. Besides, the contribution of corn bioethanol in addressing the global warming issues is very modest while having a small positive net energy balance; i.e. the energy return on investment (EROI) of corn bioethanol is low (=1.2-1.6) compared to oil (=9) [6]. The emerging challenges for 1st generation of biofuel industries that utilizes corn and soybean as a major feed stock for biofuel production motivated the search for non-food and more efficient energy feedstocks like jatropha, lignocellulosic biomass and algae. Among them, cellulosic matter will be the major feedstock for second generation biofuel, since it exhibits much higher yield per hectare in comparison with sugar or starch crops [6]. As a result, cellulosic biomass can potentially yield higher land fuel (135 GJ/ha) than corn kernel (85 GJ/ha) and soy (18 GJ/ha) [15]. Moreover, significantly higher carbon sequestration is another advantage of the use of cellulosic biomass in biofuel production compared to the first generation biofuel crops [6].

This biomass-biofuel conversion can be performed under three major classes and they are (i) conversion of renewable polysaccharides into sugar molecules and their effective fermentation into ethanol, (ii) syngas production and their bio/chemical conversion into alcohols and (iii) production of bio-oil though fast-pyrolysis and their upgrade into transportation fuels. Considering the lignocellulosic feedstock as the biofuel precursor, it is crucial to create the necessary infrastructure in many levels from biomass to biofuel production; agriculture–technology–policy. The new utilization of biomass would largely affect the agriculture sector and necessitates effective actions to ease the adaptation process. Biofuel production uses land which keeps it from food production and environmental preservation. Other issues might be considered; soil erosion may worsen by expanding the biomass production, reduction of environmental land affects biodiversity and more pesticides and fertilizers may be used. Thus, the sustainability of biofuel is not achieved solely by a positive net energy balance [6]. In spite of all raised issues, it is important to bear in mind that biofuel still offers its advantages even if it has a small contribution compared to fossil fuels. In this regard, the two main challenges in biomass production can be (a) developing crops with suitable physical and chemical traits for biofuel production and (b) increasing biomass yields (double or more) [7]. How to put these two different strands into an integrated production strategy is important and brings new research topics into the whole agriculture picture. The outcome of such productivity-enhancing innovations, research and development motivated by biofuel can be such that by 2050, the whole world population could be supplied enough diet while less cropland is used than today [6]. The respective biorefinery operation also needs improvements so that sugars can be produced from cellulosic biomass and fermented economically feasible and able to compete with production from corn and sugar. These include improvement in lignocellulosic pretreatment, reduction in enzyme (cellulase) cost, both cellulase production and ethanol fermentation by using modified microorganisms [7].

In general, the growth of biofuel industry consists of (i) increased production capacity and (ii) successful transformation of industrial technology from discrete batch method (small-scale) into continuous flow method (larger-scale) [16]. In most of the small scale manufacturing, the industries do not have the practice of collecting coproducts, thus they run with increased operating costs. Hence, larger-scale industries are keen in capturing their coproducts in order to reuse them in the production process, which results in the reduction of operating cost significantly. Thus, value-added processing may serve as a viable alternative that not only reduces the impact on the environment, but also generates additional revenue source for biofuel plants [17]. Sustainable bioeconomy road map that integrates renewable resources, biofuel production/ utilization and the value-addition to the respective coproducts is shown schematically in Figure 1. In recent years, biofuel coproducts have been utilised for the fabrication of various chemicals for diversified applications and used as the filler/reinforcement for polymer blends as well as composites. The emerging opportunities for the biofuel coproducts in biomaterials (polymers/ composites) applications make successive transformation of coproducts to renewable feedstock with economic benefits. Capitalizing this transformation enhances the economic viability and also the sustainability of biofuel industries. Thus, this chapter summarizes the various aspects in biomaterial applications of the biofuel coproducts and their role in sustainable bioeconomy.

2. Current status of coproducts from biofuel industries

2.1. The global biofuel industry status

The biofuel industry has been growing rapidly during recent years and continues to expand for the next decade. Such expansion is basically driven by renewable energy goals and different policy supports as for example use mandates, tax relief, fuel quality specifications and investment capacities in leading producing countries [18]. Based on the projection reported by

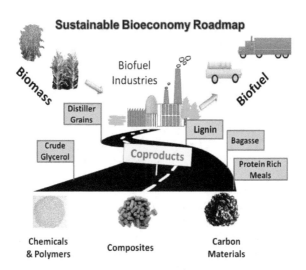

Figure 1. Sustainable bioeconomy roadmap.

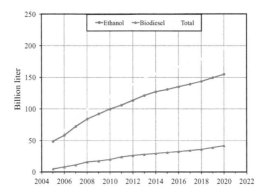

Figure 2. Development of global biofuel industry (drawn from data reported in [19, 21]).

the OECD (organization for economic co-operation and development) – FAO (food and agriculture organization of the united nations) Agricultural Outlook for the 2011-2020 period, the global bioethanol industry will be growing almost 68% from an average of 92 billion liter in the 2008-2010 period to 155 billion liter in 2020 (Figure 2) [19]. In this regard, coarse grains and sugarcane are going to remain the major precursors in bioethanol production and in 2020 they are expected to account for 78% of bioethanol feedstock (Figure 3) although this value was 81% on average during the 2008-2010 period [20]. The large scale production of cellulosic ethanol is still not achieved and under research and development. Therefore, it is expected to expand in the latter projected years reaching up to more than 4 billion liter in 2020. This is far less than the respective value for the first generation ethanol. The rest feedstocks include wheat (3.9%), molasse (3.2%), non-agricultural feedstock (2.6%), sugar beet (2%) and other (5.8%) [20].

Similar trend has been presented for biodiesel as illustrated in Figure 2 [21]. The growth in this industry in 2020 is projected to be almost 138% compared to 2008-2010 period on average; an increase from 17.6 to 41.9 billion liter. Vegetable oils will contribute more than 78% as the main feedstocks for biodiesel production. The application of non-food oils such as jatropha in biodiesel production still remains very less as compared with the contribution of vegetable oils such as soybean and palm oil. Feedstocks other than edible oils in biodiesel production include non-agriculture feeds (12.3%), biomass-based (6%) and jatropha oil (3.2%) (Figure 3) [22]. The huge impact of such expansion in biofuel industry on the respective coproducts is incontestable. Based on the feedstock share in the biofuel production by 2020, the major coproducts of different sectors of the biofuel industry can be listed as dried distillers' grains with solubles (DDGS) from dry mill corn ethanol, corn gluten meal and corn gluten feed from wet mill corn ethanol, bagasse from sugarcane ethanol, lignin from second generation lignocellulosic ethanol, and soy meal and crude glycerol from biodiesel. More focus on these coproducts will be dedicated in the following sections.

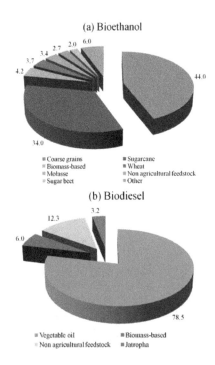

Figure 3. The global biofuel production by feedstocks contribution (%) in 2020 (redrawn from data reported in [20, 22]).

2.2. Biofuel coproducts

2.2.1. Corn bioethanol

Starch-based ethanol can be obtained from corn, wheat, barley, sorghum or any other starchy grain by fermentation. However, due to highly fermentable starch content, corn is the main feedstock for ethanol production by fermentation and accounts for 98% of all starch-based ethanol feedstocks [1]. Bioethanol from corn is produced in both dry mill and wet mill plants each of which producing specific coproducts as described below.

2.2.1.1. Dry mill

In a dry mill, ethanol is produced from corn after several steps including grinding, slurrying, cooking, liquefaction, saccharification, fermentation and distillation. Further steps are implemented to separate coproducts such as centrifugation, evaporation and drying. From the original corn mass before processing, approximately one third results in carbon dioxide during fermentation, one third is converted into ethanol and the residue are nonfermentable components in the form of different coproducts namely dried distillers' grains with solubles (DDGS), dried distillers' grains (DDG), wet distillers' grains (WDG) and condensed distillers' solubles (CDS). The coproducts are mainly dried and sold as dried distillers' grains with solubles (DDGS). This way, it is possible to store the coproduct for a longer time or ship it to far distances with less probability of fungi attack. A smaller part of the coproducts are shipped wet locally for immediate usage [23-24]. Distillers' grains have been traditionally using as animal feed due to its nutritious value as shown in Table 1 [25-31]. At the end of ethanol production process, when most of the grain's starch portion is fermented, there is an increase of 3 to 4 times in other components of the grain including protein, lipid and fibre over that contained in the unconverted whole grains [27].

Several attempts and studies have been published on distillers' grains application as animal feed in many different species such as dairy cattle [32], beef cattle [33], swine [34], broiler [35], laying hen [36], turkey [37], lamb [38], catfish [39], tilapia [40], trout [41] and prawn [42]. However, four major livestock species to which distillers' grains is practically fed are beef cattle, dairy cattle, swine and poultry [43-45]. Renewable Fuel Association (RFA) reports the distillers' grains consumption in 2009 in different species at approximately 39% for dairy cattle, 38% for beef cattle, 15% for swine, 7% for poultry and 1% for other species [24]. The important question here is whether the increasing supply of distillers' grains can be totally consumed by animals or the supply far exceeds its demand as feed. According to Hoffman and Baker [44] and Tokgoz et al. [46] the potential domestic and export use of distillers' grains in U.S. exceeds its production and the U.S. beef sector is the dominant user of distillers' grains. However, such opinions need precise consideration with respect to the fact that incorporation of distillers' grains within animal diets exhibits some limitations. Since distillers' grains are highly concentrated in terms of nutritious content, it should be included as a part of animal feed. In this regard, Canadian Food Inspection Agency (CFIA) has set out the policy for the maximum inclusion rates of distillers' grains in the feed of different species [47]. For example, the inclusion rates of distillers' grains in the diet of beef cattle and swine must not exceed 50% on a dry basis. This suggests that continuing the use of distillers' grains in animal diet in order to keep the track with its increasing supply from ethanol production can only come true if the number of consumer animals is also increasing. In other words, finding new value-added usages for distillers' grains within feed sector should also be considered in the future.

	DDGS [25-27]	CGM [28]	CGF [28]	SM [29]	CM [29-30]	JM [31]
Dry Matter (%)	88.8-91.1	90	87-90	NA	91.5	NA
Protein (% DM)	24.7-32.8	60	18-22	53.5-54.1	38.3	55.7-63.8
Fat (% DM)	11.0-16.3	2.5	2-5	1.4-2.3	3.6	0.8-1.5
Acid Detergent Fiber (ADF) (% DM)	12.4-15.2	5	13	7.2-10.2	17.5	5.6-7.0
Neutral Detergent Fiber (NDF) (% DM)	46.1-51.6	NA	35	9.6-13.8	21.5	8.1-9.1
Ash (% DM)	4.2-12.0	1.8	6.5-7.5	7.2-8.1	8.1-8.6	9.6-10.4

DDGS: dried distillers' grains with solubles, CGM: corn gluten meal, CGF: corn gluten feed, SM: soybean meal, CM: canola meal, JM: jatropha meal, NA: not available

Table 1. Composition of different biofuel coproducts

It is worth to note that the U.S. dried distillers' grains with solubles (DDGS) exports already doubled in 2009 compared to 2008 and U.S. has managed to increase its export of DDGS in 2010 by 60% compared to 2009 [48]. This may suggest that there is an excess of DDGS supply over its consumption in animal feed sector in the United States. Moreover, it should be carefully examined how the revenue from distillers' grains sale as feed, returning to the biofuel industry will economically help the ethanol industry. For corn biofuel industry to stay viable, the applications of its coproduct, distillers' grains, need to be expanded [23]. Consequently, the new outlets of distillers' grains may add value to it and create revenue for the corn ethanol biofuel. Such new usages can be value-added animal [23, 49] and human food [50], burning [51-52], extraction of zein [53], cellulose [54] and oil [55-56] from distillers' grains, and biobased filler for polymer composites, which is going to be discussed more later on.

2.2.1.2. Wet mill

Wet milling is a corn processing process in which the produced corn starch can be fermented into ethanol. Thus, ethanol is not the only product of a wet mill. In the beginning, the feedstock goes through a steeping step that soaks it in warm water containing small quantities of dissolved sulfur dioxide for almost 40 hours. This step facilitates the separation of the grain components. The different processes will be then applied such as grinding, screening, germ separation, oil refining, starch-gluten separation, drying, fermentation and syrup refining. Other products of a wet mill plant produced along with ethanol in these processes include starch, corn oil, high fructose corn syrup (HFCS) and glucose/dextrose. The coproducts of different steps are corn gluten feed, corn gluten meal, corn germ meal and corn steep liquor [28]. Corn gluten feed and corn gluten meal are the major feeds for livestock produced in a wet mill. As compared to the produced ethanol in a wet mill, corn gluten feed and corn gluten meal are produced almost as much as 70% and 17% of the mass of the produced ethanol, respectively [44]. The composition of the wet mill coproducts are presented in Table 1. Similar to distillers' grains, these coproducts account for a good source of nutritious components such as protein and fiber for feed applications. The value of these coproducts for animal feed has been realized for many years now and they are used as the feed for a wide range of animals including beef cattle [57], calf and lamb [58], dairy cattle [59], poultry [60], swine [61], pet [62] and fish [63]. Also, it has been reported that corn gluten meal can be used as the pre-emergence weed control (to control weeds before the weed seeds germinate) [64] and has been regulated by US Environmental Protection Agency (EPA) [65].

2.2.1.3. Corn ethanol coproducts production

Although there is a lack of comprehensive statistics on global production of corn bioethanol coproducts, a general insight can be obtained considering the production of these coproducts in the United States during last 10 years. This may be reasonable since U.S. is the largest ethanol producer globally with more than 50% contribution in 2007 and 2008 [8]. As reported by the Renewable Fuels Association (RFA) (Figure 4), the production of distillers' grains, corn gluten feed and corn gluten meal in the U.S. have shifted totally more than 10 times from 3.1 million tonnes in 2001 to 32.5 million tonnes in 2010 [24].

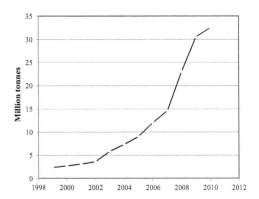

Figure 4. Production of ethanol biorefineries coproducts in US including distillers grains, corn gluten feed and corn gluten meal (drawn from data reported in [24]).

2.2.2. Sugarcane bioethanol

In a sugar mill, the crushed sugarcane is washed to go for juice extraction. The resulted juice can be used for sugar as well as ethanol production. Bagasse is the lignocellulosic coproduct after the sugarcane is crushed for juice extraction [66]. Approximately, it consists of 50%

cellulose, 25% hemicellulose and 25% lignin [67]. Bagasse has been widely used as the fuel for generating electricity. One metric ton of bagasse containing 50% moisture will produce heat equivalent to that from 0.333 tons of fuel oil [68]. This coproduct has been considered for such purpose in different countries such as Zimbabwe [69], Nicaragua [70] and Brazil [71]. Another large utilization of bagasse is in paper and pulp industry. This was patented in 1981 [72] and found huge application in many places such as India as early as 1990 [73]. The particleboard production is another industrial utilization of this biomass [74]. Bagasse has been also used in composting to a limited extent [68] and use of fungal strains on bagasse has been reported to produce compost with low pH and high soluble phosphorus [75]. Fermentation of bagasse using mold cultures was also considered to produce animal feed [76].

As a lignocellulosic material, bagasse has the potential of a feedstock for biofuel production either by gasification or hydrolysis method. In this context, still the biofuel production from bagasse via gasification has not been reported. However, as an alternative method competitive to the direct combustion of bagasse, gasification using a two-stage reactor has been proposed to be economically viable and more efficient [77]. Also, studies have been conducted in order to improve the bagasse gasification as far as retention and separation of alkali compound is concerned during the process. Considering the lignocellulosic ethanol production from bagasse, several investigations has been published on liquid hot water, steam pretreatment [78] and acid hydrolysis of it [79] as well as simultaneous saccharification and co-fermentation (SSCF) method [80]. As published in 2004, about 180 MT of dry sugarcane bagasse is produced globally and can be utilized to produce about 51 GL of bioethanol [81].

The expansion in production of the sugar-based ethanol is one of the key factors affecting the bagasse production. The sugarcane harvest of Brazil, the global leader of sugar-based ethanol, has shifted upward approximately 45% during recent five years from 425.4 MT in 2006-07 to 620.4 MT in 2010-11 (Figure 5) [82]. The ethanol production in Brazil generally shows a similar trend. To produce one liter of ethanol, 12.5 Kg of sugarcane is required. The weight of the produced bagasse is about 30% of the weight of sugarcane used for sugar or ethanol production [66]. Therefore, the bagasse production of Brazil in 2010-11 can be estimated as more than 180 MT, almost equal to global bagasse production before 2004.

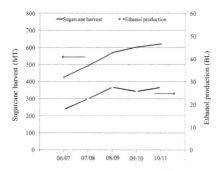

Figure 5. Brazilian sugarcane harvest and ethanol production (drawn from data reported in [82]).

2.2.3. Lignocellulosic bioethanol

In a second generation bioethanol plant, lignocellulosic polysaccharides (cellulose and hemicellulose) are broken down into monosaccharides (hexose and pentose) to be further fermented into ethanol. This includes a sequence of processes such as pretreatment, hydrolysis (enzymatically or chemically), fermentation and purification. At the end, the residual from the original lignocellulosic biomass is the coproduct mainly in the form of lignin [83]. The amount and quality of the produced lignin depends on the original lignocellulosic matter and the process. Typically, lignocelluloses contain 10–30 % lignin, which depends on various factors such as nature of biomasses, growth as well as isolation process [84]. As we know, lignin is a polymer that exists in the cell walls of plants, which is available in nature next to cellulose. The role of lignin in plant is to save them from compression, impact and bending. In addition to that, the major role of lignin extends to prevent the plant tissues from various kinds of naturally occurring microorganisms [85].

Chemically the polymeric structure of lignin is highly complicated and consists of three different monomer units (Figure 6) and they are called as p-coumaryl alcohol, coniferyl alcohol and sinapyl alcohol [86-87]. In addition to plant varieties, the lignin extraction process (known to be delignification) play a major role in the determination of local structure. Thus, the chemical structure of lignin extracted from plant biomass is never similar as exist in plant. There is a possibility for the alteration of monomer arrangement in the lignin structure. Hence, the lignin in plants, called as "natural lignin", is termed as "technical lignin" after isolation. These technical lignins can be further classified into three classes based on the domination of monomer units and they are [88]: (i) Softwood lignin: dominated with coniferyl monomer units, (ii) Hard wood lignin: combinations of equal quantities of guaiacyl and syringyl monomer units and (iii) Grass lignin: equally formulated with all three monomers of coniferyl, sinapyl and p-coumaryl.

The sources for lignin production can be divided into two major categories and they are (i) paper and (ii) bioethanol industries. They use different types of de-lignification process thus the lignin from paper industries and the lignin from bioethanol industries are not similar. Paper industries adopt Kraft, Sulphite, and Soda pulping processes to remove lignin from biomass, where as lignocellulosic ethanol industries, prefer to go with organosolv, steam explosion, dilute acid as well as ammonia fibre explosion for the removal of lignin [89]. Due to the tough food-versus-fuel concerns, non-food crops such as switchgrass, miscanthus and sugarcane bagasse become an effective feedstock for ethanol production and those related biofuel are named as second generation biofuels [90]. As the global demand for biofuel continues to grow, there will be emerging opportunities for lignocellulosic ethanol industries, which is expected to create a huge amount of lignin and it is predicted to be ~225 million tons by 2030 [85]. The challenge is to dispose them effectively that includes using them as a feedstock for energy products as well as for the fabrication of various chemicals and materials [91]. In common practice, lignin has been used for energy fuel, which is also the easiest way of disposal. In other hand, value-added uses of lignin can give economic return to the lignocellulosic ethanol industries and can improve their sustainability.

Only 2% of the lignin, which is produced from various sources, has been used as the feedstock for various chemicals including phenol, terephthalic acid, benzene, xylene, toluene, etc. [89] In addition to these, lignin has also been used in fertilizer, wood adhesives, surfactants, and also some kind of coloring agents [89]. Recently, lignin has been included as filler/reinforcing agent for blends and composites in both thermoplastic as well as thermoset platforms [85, 92-94]. In addition to that, lignin is found to be a suitable renewable carbon source for the synthesis of carbon materials [95]. Lignin has been widely exploited for the fabrication of activated carbon for various purposes including hydrogen storage, waste water removal and energy storage/conversion. Especially synthesizing nanostructured carbon materials from renewable resource-based lignin receives a great scientific interest due to the unique morphology as well as their physicochemical properties. As the global demand grows for the carbon fibre composites, there is a huge demand for the low cost carbon fibres. Lignin-based carbon fibres can substitute polyacrylonitrile (PAN)-based carbon fibre, hence the opportunity for lignin as a successful feedstock for carbon fibre is in near future [96]. This will be possible by understanding the basics of lignin chemistry and their application for fibre fabrication.

2.2.4. Biodiesel

2.2.4.1. Proteineous meal

The major biodiesel feedstocks are vegetable oils which are generally produced by crushing oil seeds, leaving significant quantity of proteineous meals as coproducts. The global consumption of proteineous meal in 2011/12 as reported by USDA Economic Research Services .[97] is depicted in Figure 7 indicating soybean as the predominant crop producing proteineous meal with 67 % contribution. According to the FAOSTAT database [98], US, Brazil and Argentina were the global premiers of soybean production in 2010 with 35, 26 and 20% contribution, respectively. The next largest producer is China with almost 6% share in the global soybean production. In this context, biodiesel production of the US, as for example, has increased, during 2006-11 period, 340% from 250 to 1100 millions of gallons [99], which promoted the proteineous meal production. Soybean meal is traditionally used as a filler in animal feed including poultry, swine, beef, dairy, pet and other animals due to its concentrated protein content (Table 1).

Other examples of plant-based feedstock potentially suitable for oil extraction and biodiesel production can be listed as canola and linseed [100], palm [101], karanja [102] and jatropha [103]. Jatropha is a non-edible seed from a large shrub commonly found throughout most of the tropical and subtropical regions of the world. As shown previously in section 2.1, jatropha is projected to have a 3.2 % contribution in biodiesel production by 2020. The growing utilization of plant-based feedstocks other than soybean meal in biodiesel production also brings new streams of proteineous meal as coproducts.

2.2.4.2. Crude glycerol

Biodiesel is chemically known as methyl esters, which is produced through transestrification reaction by reacting a vegetable oil or animal fat with an alcohol under a strong base catalysis environment [104]. Along with biodiesel, such transestrification reaction produces significant

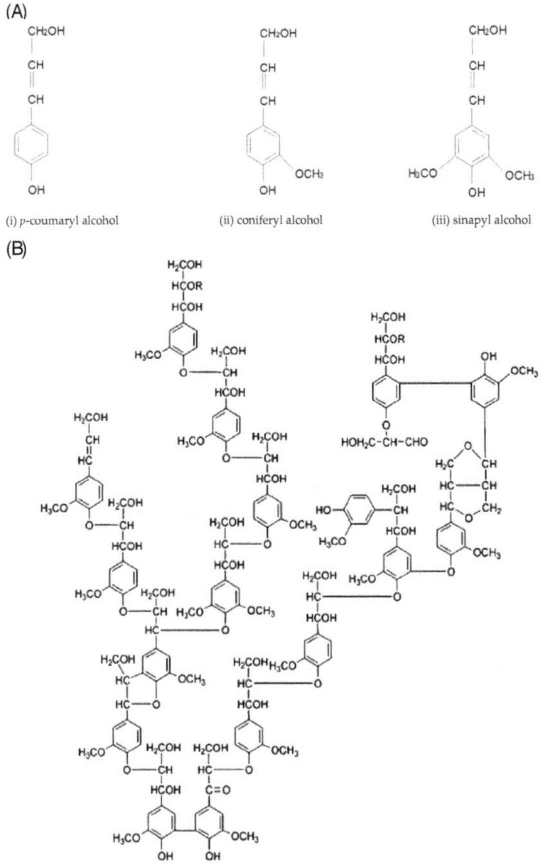

Figure 6. (A) Phenolic precursors that form the lignin and (B) Chemical structure of lignin (reprinted the figure with permission) [87].

quantity of glycerol (also called as glycerin), which is normally collected with other ingredients such as catalysis, water and unreacted alcohol and it is termed as crude glycerol [105]. Normally biodiesel industries utilize excess amount of methanol as required for the completion of reaction, which leaves unreacted methanol to the glycerol after the reaction. With every 3 gallons of biodiesel, 1kg of crude glycerol is produced and it shows very low value because of its impurity [106]. As the global biodiesel production increases exponentially, the resulting crude glycerol is extensively high and become issues due to their disposal or effective utilization. On the other hand, pure glycerol has found a wide range of applications that includes food, cosmetics, and drugs. In order to upgrade the crude glycerol to those high end applications, it should undergo various purification stages such as bleaching, deodoring, and ion exchange. Normally, this is not affordable nor economically feasible for most of the small/

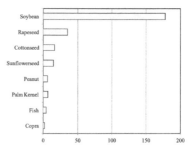

Figure 7. World protein meal consumption (million tonnes) in 2011 (drawn from data reported in [97]).

medium ranged industries. Hence, it is necessary to investigate the value-added uses of crude glycerol in various applications.

Prior to that, it is necessary to understand the relationship between the oil feedstock and the crude glycerol. Thompson and He [105] performed a research on the characterization of crude glycerol samples from various feedstocks. Their research shows that the compositions of different crude glycerol are highly varying with their feedstocks. This creates the challenge to adopt a universal protocol to fabricate value-added products from crude glycerol from various feedstocks. Crude glycerol has been used to produce various products including 1,3/1,2-propanediol, dihydroxyacetones, polyesters and hydrogen [107]. Mu et al. [108] reported the synthesis of 1,3-propanediol using crude glycerol produced during biodiesel preparation through fermentation process using Klebsiella pneumonia. They used the crude glycerol obtained during soybean oil-based biodiesel production employing alkali catalysis. They ultimately compared the product of 1,3-propanediol obtained from pure glycerol and found that they are similar to each other [108]. Soares et al. [109] demonstrated the generation of synthesis gas or syngas (hydrogen and carbon monoxide) from glycerol at very low temperature between 225-300 °C employing a Pd-based catalyst. Further it can be converted into fuels/chemicals by Fischer–Tropsch methanol synthesis. They also suggest this process for the effective utilization of various crude glycerol feedstocks for the fabrication of high value fuels/chemicals. Mothes et al. [110] reported the synthesis of poly (3-hydroxybutyrate), PHB, using crude glycerol (rape seed oil-based) as the feedstock via biotechnological process employing Paracoccus denitrificans and Cupriavidus necator microbes. They compared the properties of the synthesized PHBs from two different feedstocks and found that the properties are very similar. Zhou et al. [111] reviewed the chemo-selective oxidation of crude glycerol into various products such as glyceric acid, hydroxypyruvic acid and mesooxalic acid which can be used as precursor for various fine chemicals and polymeric materials. These reports indicate the emerging opportunities for crude glycerol for various applications including chemical, fuel and materials.

2.3. Sustainability through value addition

Due to the uncertainty in long term availability of fossil fuels and their continuous threat to environment through greenhouse gas emission, there is a drive across the globe towards the exploration of various biorefinery systems. Production of biofuel creates impact in utilization of biomass, replacement of possible extend of gasoline, reduction of greenhouse gas emission and the creation of significant amount of coproducts [112]. Biofuel production from biomass tends to strengthen the entire value chain (farming community-biofuel industries-consumers) and claims as the probable sustainable alternate for the conventional fossil fuel systems. Biomass is the sustainable feedstock for biofuel industries and biofuel provides ecological safety towards sustainable transportation, however the emerging concern is about the co/by-products. If they create challenging environmental issues in disposal, sustainability of this technology is challenged. Thus, value addition to these biofuel coproducts plays key role for the sustainability of biofuel technology in long term perspective [113]. With this understanding, the ultimate aim of biorefinery is focused to satisfy the conceptual "triple bottom line" of sustainability that includes (i) economic development; commercial value for biomasses, biofuels and coproducts, (ii) social development; appearance of new manufacturing sectors as well as creation of rural job opportunities, and (iii) environmental/resource sustainability; greenhouse gas reduction and eco-friendly green products [114]. In order to ensure the sustainability of biofuel technology it is essential to address various issues including (i) "food vs. fuel" due to the usage of edible resource for biofuel production, (ii) resource availability/management; effective utilization of land/water resources, (iii) environmental impact; issues related to land/water quality retention, conversion of grasslands/forests to agricultural fields and the efficient disposal/utilization of biofuel coproducts and (iv) validated measures: policy making and certification/standardization [115]. Scale-up activities of biofuel production is essential due to the increasing demand for the substitution of fossil fuel. However, it is significantly controlled by various factors such as effective land usage for the larger biomass generation, water availability for agricultural forming, retention of soil quality, environmental impact of biofuel coproducts and labor market shift towards biorefinery [116-117]. The successive transformation of biofuel production from conventional to second generation effectively addresses the issues related to water consumption. Lignocellulosic ethanol industries utilise perennial crops such as miscanthus and switch grass, which grows on marginal land and consume very less water. The challenge is towards biofuel coproducts. Failure of handling these large quantity coproducts will ultimately create serious environmental issues. These emerging technologies related to the effective utilization of biofuel coproducts that holds significant quantity of renewable content significantly substitute/ replace petroleum-based products and helps in reduction of greenhouse gas emission.

3. Value-added biomaterials from biofuel coproducts

3.1. Distillers' grains

3.1.1. As biofiller in producing polymeric biocomposites

The low cost of distillers' grains (DDG and DDGS) is a key incentive for researchers to utilize them as biobased fillers in manufacturing polymer composites. Also, addition of DDG(S) to the polymer matrix can result in improved stiffness as long as proper treatments and processing aspects are taken into account. It is less than ten years that DDGS-containing biocomposites has been reported in the literature. In this regard, the very first produced DDGS-containing composites exhibited low mechanical properties so that the utilization of DDG(S) as a biofiller in composite materials seemed to be not worthy in the beginning. However, recent works project a better future for DDG(S)-based biocomposites.

3.1.1.1. DDGS-polyolefin biocomposites

Polyolefins such as polypropylene (PP) and polyethylene (PE) were the first polymers compounded with DDGS up to 30 wt% [118]. The mentioned work included a comparison of four types of biofiller such as big blue stem (BBS) grass, soybean hull, pinewood and DDGS in terms of the mechanical performance of the their biocomposites with PP and PE. The composite processing was performed in a twin screw extruder and mechanical properties tests were conducted. Generally, the studied biofillers increased the flexural and tensile moduli. However, DDGS increased the modulus not significantly in comparison with neat PE and PP. Moreover, the tensile and flexural strengths decreased drastically as a result of compounding with DDGS. In general, the authors of [118] came to a conclusion that DDGS is not a suitable biofiller because of the decreased mechanical properties of the studied DDGS/polyolefin composites. In another work, the composite of high density PE with DDGS has been produced via extrusion and injection molding technique with DDGS content of 25 wt% [119]. The effect of maleated polyethylene (MAPE) as the compatibilizer was studied. Moreover, DDGS was solvent-treated to remove its oil and polar extractables. It has been reported that the application of the MAPP with solvent-treated DDGS resulted in better tensile and flexural properties of the composite compared to the respective properties of the neat HDPE.

3.1.1.2. DDGS-polyurethane biocomposites

This is another attempt of biomaterial application of DDGS. The overall idea of such work was to utilize a tough binder between rigid particles of DDGS to create an acceptable flexible material. Thus, polyurethane prepolymer (PUP) from castor oil was used as a binder with different compositions of DDGS and the biocomposites were prepared in a two-step process; PUP and DDGS were premixed in a micro-extruder and then compression molded to the shape of sheet. The mechanical and dynamic mechanical properties characterizations showed that the produced PUP/DDGS sheet was more flexible compared to the brittle DDGS material, thus polyurethane enhanced the properties of DDGS [120].

3.1.1.3. DDGS-phenolic resin biocomposites

Tatara et al. [121-122] bonded DDGS particles, from 0 to 90 wt%, with phenolic resin via compression molding process. Mechanical properties of the blends showed a reduction in modulus, tensile strength and elongation at yield by increasing the DDGS amount. However, the researchers believed that the cost saving resulted from the addition of low-cost DDGS filler in a reasonable quantity may offset the reduction in property performance; i.e. a cost-performance balance is achieved. In this context, inclusion of 25 and 50 wt% of DDGS maintained the mechanical strength to an acceptable value. In another work, the effect of DDGS particle size and content (25 and 50%) was studied when phenolic resin-based glue and wood glue were used to produce DDGS composites [123]. Overall, the DDGS composite with resin glue showed better mechanical properties and curing uniformity compared to wood glue/DDGS composites. Also, DDGS enhanced the flexural properties such as modulus and maximum stress. Also, composites of DDGS with smaller particle size (0.7 mm) had higher mechanical properties compared to those with higher particle size (0.34 mm).

3.1.1.4. DDGS biopolymer biocomposites

During recent years, biopolymer thermoplastics such as poly(lactic acid), PLA [124], poly(butylene succinate), PBS [125], polyhydroxy(butyrate-co-valerate)/poly(butylene succinate), PHBV/PBS, blend [126] and poly(butylene adipate-co-terephathalate), PBAT [127] have been utilized to produce DDGS composites. The influence of DDGS amount from 20 to 50 wt% as well as compatibilizer in PLA/DDGS composites were investigated [124]. Drastic decrease in tensile modulus and strength was observed by increasing the wt% of DDGS when no compatibilizer was used. On the other hand, after using isocyanate type of compatibilizer, a huge improvement in tensile modulus and strength was observed in the PLA/20% DDGS composite (Figure 8). In comparison with pure PLA, the compatibilized formulation showed higher modulus and almost equal strength. In another work, the thermal degradation of DDGS was studied with considerations for biocomposite processing and it was reported that water-washing of DDGS improved the thermal stability of DDGS to the extent that its thermal decomposition was highly prevented at typical temperatures of polymer melt processing. Such improvement in thermal stability of DDGS resulted in better strength and modulus of the PBS/DDGS biocomposite with 30 wt% DDGS [125]. The effect of compatibilizer was studied in a composite of 30 wt% DDGS and PHBV/PBS blend processed in a micro-extruder/micro-injection molding machine [126]. The DDGS used in this biocomposite had a water-washing step prior to compounding with bioplastics. Using a compatibilizer (isocyanate type), the interfacial adhesion was enhanced. The optimized biopolymer/DDGS composite exhibited improved tensile modulus compared to the biopolymer matrix while having almost equal strength. The influence of DDGS on the biodegradability properties of PBAT/DDGS biocomposites has been evaluated [127]. It was observed that PBAT/DDGS biocomposite was found to be more bio-susceptible material compared to virgin PBAT and was totally biodegraded. During the biodegradation experiment DDGS domains were preferentially attacked by microorganisms and influenced the biodegradability of the PBAT matrix. The produced biocomposite showed a degree of biodegradation similar to the biodegradation rate of natural materials such as DDGS and cellulose.

3.1.2. Other biomaterial applications

DDGS as a source of protein, fiber and fat has been used to isolate these components. Xu et al. [53] implemented a novel acidic method to extract zein from DDGS which is the main protein in corn and corn coproducts such as DDGS. The resulted zein has the potential for uses in fibers, films, binders and paints applications. Their method also isolated DDGS oil during protein extraction. Other researchers have also investigated the extraction of oil from DDGS

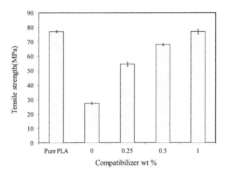

Figure 8. The effect of compatibilizer on tensile strength of the biocomposite PLA-20wt% DDGS (drawn from the data table with permission) [124].

[55, 128]. The obtained oil can be used as precursor for biodiesel production. It has been also tried to extract fiber from DDGS with physical methods such as sieving and elutriation [129] or extract cellulose chemically by sodium hydroxide solution [54]. The obtained cellulose had properties suitable for films and absorbents. The bioadhesive formulation obtained from DDGS is another biomaterial application of this coproduct [130-131]. This glue is prepared by reaction with an aqueous base solution. Urea can be also included with the base. The obtained bioadhesive is particularly useful as boxboard glue.

Recently, it has been tried to produce thermoplastics from DDGS by chemical methods such as acetylation [132] or cyanoethylation [133] with almost similar approaches. Hu et al. [133] were successful in producing highly flexible thermoplastic films from DDGS. The oil and zein protein of the DDGS were extracted first and the resultant underwent cyanoethylation using acrylonitrile. A compression molding machine was implemented to produce oil-and-zein-free DDGS films. It was observed that the produced films had much higher strength even at high elongations compared to films developed from other various biopolymers. Therefore, cyanoethylation could be a viable approach to develop bio-thermoplastics from biopolymers for applications such as packing films, extrudates and resins for composites. It has been also tried to find a novel use of dried distillers' grains (DDG) as a feedstock for bio-polyurethane preparation [134]. The procedure consist of, first, liquefaction of DDG in acidic conditions at atmospheric pressure and then reaction of hydroxyl-rich biopolyols in the liquefied DDG with methylene diphenyl diisocyanate (MDI) to form networks of cross-linked polyurethane. Thus, DDG-based bio-polyols were the precursor in this way to synthesize flexible and rigid polyurethane foams. The biodegradation tests showed that the degradation of these polyurethane foams in a 10-month period was about 12.6% most probably because of natural extracts such as proteins and fats in DDG and partially cross-linked or uncross-linked residue in the foam [134].

3.2. Bagasse

Bagasse has been used in biomaterial applications since a very long time ago. It has been used for interior panels and particleboard production. The first bagasse composition panel plant in Americas was built by Celotex, Louisiana, in 1920. Since then, more than 20 bagasse particleboard plants have been built throughout the world [135]. However, recent characterization of bagasse fiber for its chemical, physical and mechanical properties indicates that the potential of this coproduct of sugar and biofuel industries is much more than its applications in interior and structural components [136]. Bagasse is mostly burnt to generate energy for the sugar industry itself. Considering the fact that for such purpose almost 50% of the bagasse's production is enough [137], it is necessary to develop new uses for these fibers to implement the rest 50 % and reduce their environmental impact. Moreover, the burning of bagasse fiber is also a matter of concern as far as atmospheric pollution because of smoke, soot and ash is concerned [138]. Chemical composition as well as physical and mechanical properties of bagasse fiber are presented in Table 2 [136, 139-142]. Bagasse fiber consists of structural components such as cellulose and hemicellulose that can provide stiffness and rigidity to the polymers and enhance their engineering applications. Besides, bagasse exhibits a porous cellular structure with a hollow cavity called lumen existing in unit cell of the fibers. Therefore, the bulk density of bagasse fiber is lower than other natural fibers and bagasse fibers can act more effectively as thermal and acoustic insulators [142]. For example, the densities of kenaf and banana fibers are 749 kg m^{-3} [140] and 1350-1500 kg m^{-3} [139], respectively, which are higher than that of bagasse (344-492 kg m^{-3} [139-140]). Also, cellulosic fibers such as bagasse with low Young's modulus can act as useful crack growth inhibitors [143].

	Properties		Ref.
	Cellulose (%)	32.0-55.2	[139]
	Hemicellulose (%)	16.8-32.0	[139]
Chemical composition	Lignin (%)	19.0-25.3	[139]
	Ash (%)	1.1-4.3	[139]
	Extracts (%)	0.7-3.5	[139]
	Density (kg m^{-3})	344-492	[139-140]
Physical properties	Diameter (μm)	394-490	[136, 140-141]
	Moisture content (%)	52.2	[136]
	Water absorption (%)	235	[136]
	Tensile strength (MPa)	29.6-96.2	[136, 141-142]
Mechanical properties	Tensile modulus (GPa)	4.5-6.4	[140, 142]
	Elongation at break (%)	4.0	[142]

Table 2. Chemical composition, physical and mechanical properties of bagasse fiber

3.2.1. Bagasse particleboards

Bagasse particleboards generally consist of bagasse fibers bound together with either an organic or inorganic binder. The organic binders are mostly a phenolic or polyester thermoset resins and the board is produced by compression molding under high pressure and temperature. Different inorganic binders such as cement, gypsum and calcined magnesite can also be used to produce bagasse boards [144-146]. Besides, binderless bagasse particleboards have been produced and patented in 1986 which can simplify the manufacturing process and reduce production cost since the blending operation and equipment are eliminated [147]. In this regards, different processing techniques such as hot pressing [148] and steam-injection pressing [74] have been conducted.

3.2.2. Bagasse-thermoset biocomposites

Phenolic resins are the major thermosets used for bagasse particleboards and several studies have been published on using resol [137], Novolac [149], lignophenolic [150] and other phenolic resins [143, 151] with bagasse fiber. Zárate et al. [137] studied the effect of fiber volume fraction on the density and flexural properties of composites from resol and several fibers including bagasse. They compared the efficiency criterion for mechanical performance, which relates the strength and stiffness with density, of the composites with those of typical structural materials including aluminum, magnesium, polyethylene and steel. Based on this comparison, it was concluded that the stiff composite materials produced from bagasse fibers and resol matrix are better compared to typical structural materials such as steel [137]. The effect of maleic anhydride (MA) treatment of bagasse fiber on properties of its composite with Novolac has been studied [149]. It has been reported that the composites with MA treated fibers had a hardness of 2–3 times more than that of the untreated bagasse composite and MA treatment reduced water and steam absorption of the fibers. Paiva and Frollini [150] extracted lignin from sugarcane bagasse by the organosolv process and used it as a partial substitute of phenol in resole phenolic matrices to produce bagasse-lignophenolic composite by compression molding. They observed improvement in the impact strength when sugarcane bagasse was used, but no improvement was found as a result of fiber treatments such as mercerization and esterification.

Unsaturated polyesters are another family of thermoset resins used for bagasse-based composite purposes. The effect of fiber size, its surface quality and the compression molding parameters on the flexural properties of composites from polyester and chopped bagasse fiber has been investigated. It was found that composites produced with bagasse particle size of less than 2 mm, and pre-treated for the extraction of sugar and alcohol exhibited the highest mechanical performance [138]. The effect of chemical treatments using sodium hydroxide and acrylic acid on the properties of bagasse-polyester composites has been studied. The treatments resulted in the better interaction between fiber and matrix as well as lower water absorption than composites with untreated fiber [142].

3.2.3. Bagasse-thermoplastic biocomposites

Bagasse has been used as reinforcing filler in different thermoplastic matrices such as poly(ethylene-co-vinyl acetate) (or EVA) [152-153], polyolefins [154-155] and starch-based biodegradable polyester [140-141, 156-157]. The effect of cultivar type and surface cleaning of the bagasse fiber on the tensile properties of the bagasse-EVA composites have been investigated [152]. The results suggested that blends of bagasse from various cultivars can be used for commercial applications of these composites. Also, the surface cleaning of the bagasse obtained from sugar mill was good enough to use the bagasse without further surface treatment. Another study on the impact behavior of the bagasse-EVA composites showed that the mechanical performance of this type of composites could be tailored by varying the bagasse volume fraction in order to reproduce the behavior of wood-based particleboards [153]. Luz et al. [154] explored the efficiency of two different processing methods, injection molding and compression molding, to produce bagasse-polypropylene (PP) composite. They found that the injection molding under vacuum process was more efficient and created homogeneous distribution of fibers without blisters. It was observed that bagasse incorporation into PP improved the flexural modulus. High density polyethylene (HDPE) was used as the matrix for incorporation of cellulose obtained from bagasse [155]. It has been reported that modification of bagasse cellulose with zirconium oxychloride helped in improving the tensile strength of the biocomposite.

Bagasse fibers have been used to produce biocomposites from bagasse and biodegradable corn starch-based polyester which is reported as a blend of starch and polycaprolactone (PCL). The effects of volume fraction and fiber length were investigated and an optimum value for both factor were reported beyond which the decrease in mechanical performance was observed [140]. Also, it was reported that after alkali treatment of the bagasse fibers the improvement in fibre–matrix adhesion occurred that resulted in enhancement of mechanical properties [141]. Moreover, incorporation of bagasse fiber into the polyester matrix improved tensile as well as impact strength. Acetylated starch has been reinforced with bagasse fiber [156-157]. The matrix in that case was a blend of starch, PCL and glycerol. It was observed that incorporation of alkali-treated bagasse fiber up to 15 wt% increased the tensile strength while it decreased when bagasse content was more than this value. Also, the water absorption of the composite was improved as the bagasse content increased due to hydrophobic nature of bagasse compared to acetylated starch.

3.2.4. Other biomaterial applications

It has been also tried to convert bagasse fibers into a thermo-formable material through esterification [158-159]. This has been done without any solvent using succinic anhydride followed by hot-pressing to produce the test samples. The thermoplasticization of the esterified fibers was proven to occur by scanning electron microscopy. By studying the effect of pressing parameters, it has been found that that de-esterification and hemicellulose degradation could occur on certain pressing conditions. It has been claimed that the mechanical properties of the produced composite could be superior to the standard properties of conventional high density wood particleboards.

3.3. Lignin

Biomaterials applications of lignin for the fabrication of polymers, blends and their reinforced composites are highly motivated due to the following reasons (i) abundance/occurrence in nature, (ii) phenolic chemical structure and the possibility of chemical modification and (iii) eco friendliness and reduced carbon footprint [160]. The challenge to use lignin as the materials feedstock is their complicated chemical structure as well as molecular weight, which are highly dependent on the lignin extracting process and also their sources [161]. The biomaterials application of lignin is vast and it has been blended/ reinforced with a wide range of polymeric systems such as thermoplastics, thermosets and elastomers as a renewable low cost filler [162]. Utilization of lignin as the raw material for the fabrication carbonaceous materials such as activated carbon and carbon fibres has a long history, however the creation of nanostructured materials such as carbon nanoparticles/nanofibres are quite new, which has a huge commercial potential [95, 163]. The emerging opportunities for lignin in these areas are summarized in this section.

3.3.1. Polymeric blends and composites from lignin .

3.3.1.1. Lignin in thermoplastics

In polymers, lignins have been used as low-cost fillers aiming to retain their mechanical properties. Nitz et al. [164] reported the influence of various types of lignin reinforcement with the thermoplastics on their mechanical properties. Their results indicate that they are able to incorporate ~40 wt% lignin in to polyamide 11 (PA11), polyester (Ecoflex®) and polyestera-mide (BAK®) systems without impairing their mechanical properties [164]. Generally lignin shows high cross-linking/intramolecular interactions, which limits their application in solid material systems. This can be overcome through polymer blending; however, achieving miscibility is very essential to develop a material system with superior properties [165]. This is possible in lignin-based blends by manipulating the chemistry of hydrogen bonding between the OH groups and interacting sites of polymers, either polar or semi polar [165]. Moreover, the hydrogen bond with a polymer varies with lignin to lignin since the monomer combinations of the lignins are unique [166].

Lignin–thermoplastic blends can be classified into two categories and they are (i) lignin – petro-based polymer blends and (ii) lignin–renewable resources based polymer blends. Blending the lignin with polyethylene and polypropylene is well known [166-169]. Alexy et al. [166] reported the effect of lignin concentration in the fabrication of polymeric blend with PP and PE. They measured the tensile strength as the measure of mechanical properties over the various lignin compositions. For both the polymer systems they identified that the mechanical properties decrease with increasing lignin content [166]. In addition to mechanical properties, Canetti et al. [169] and Mikulášová et al. [170] reported the fabrication of lignin/PP blends and investigated their thermal and biodegradable properties respectively. Poly(vinyl chloride) (PVC) is the next popular thermoplastic, which has been produced globally and exhibits a wide range of applications [171]. Raghi and coworkers [171] reported the fabrication of lignin/PVC blend and studied their mechanical/weathering properties. Their research investigation confirmed that the addition of lignin to PVC enhanced their tensile strength and not influenced their weathering behavior. Banu et al. [172] reported the fabrication of PVC/lignin blends and investigated the effect of plasticizer in their formulations. They concluded that the specific thermal and mechanical properties are feasible in some formulations with the addition of plasticizer. In addition to that, lignin/ poly(vinyl alcohol) (PVA) and lignin/ poly(ethylene oxide) (PEO) blend systems with various types of lignins are also investigated for the effective electrospinning performance [173-176]. Sahoo et al., [92] reported the fabrication of polybutylene succinate (PBS) reinforced with renewable resource-based lignin employing a melt extrusion process. They found that lignin reinforcement in PBS enhances their properties synergistically and also achieved the incorporation of high fraction of lignin of about 65%. In addition to that, they also reported the fabrication of PBS-based composite materials with the hybrid reinforcement of lignin and other natural fibre [93]. They found that the hybrid reinforcement is more beneficial over individual reinforcement for the better flexural strength.

The research on lignin-based polymer blends with renewable resource-based biopolymer is very limited. Only few publications are available in this content. Camargo et al. [177] reported the melt processing of poly(3-hydroxybutyrate-co-hydroxyvalerate) (PHBV) with lignin, in which they used the lignin isolated from sugarcane bagasse. They found that the addition of lignin to PHBV caused a reduction in their mechanical properties, which is due to the zero integration of lignin and PHBV [177]. Mousavioun et al. [178] performed the processing of poly(hydroxybutyrate) (PHB)-soda lignin blend and studied its thermal behavior. They found that the addition of soda lignin formed the miscible blend and improved their overall thermal stability. However, they have not reported their mechanical properties [178]. Vengal et al. [179] investigated the blending effect of lignin with starch and gelatin for the fabrication of biodegradable polymeric films. They found that the addition of lignin into starch can create better film with the composition of 90:10 (lignin: starch) and further increment of lignin content decreases their properties. Casetta et al. [180] fabricated the PLA and lignin blend and investigated their flame retardant behavior. They observed that the addition of lignin to PLA enhanced their flame retardant property comported to virgin PLA.

3.3.1.2. Lignin in thermosets

The lignins of different resources have been incorporated into various thermoset resins. Amorphous/heterogeneous nature and the complex structural composition of lignin result their behavior as either like a filler or like a reactive macromonomer in epoxy resin systems. Mansouri et al. [181] characterized the alkaline lignin and suggested their possible use for phenol-formaldehyde resin due to the availability of huge OH groups. Peng et al. [182] reported the fabrication of phenol-formaldehyde thermoset resin with lignin fillers and investigated their chemo-rheological properties. They found that the curing rate of the resin system decreased with increasing lignin content. Guigo et al. [183] fabricated the poly(furfuryl alcohol)/lignin composite resins and reported their monophase behavior. This indicates the reactive monomer behavior of lignin in this thermoset system. Thielemans et al. [184] investigated the effect of kraft lignin on unsaturated thermosetting resin, which was a mixture of epoxidized soybean oil and styrene, for the fabrication of natural fibre reinforced thermoset composites. They found the complete solubility of lignin into the resin system and their result on natural fibre composites indicateed the compatibilizing effect of lignin. Nonaka et al. [185] reported the fabrication of a new resin system by aqueous mixing of alkaline kraft lignin with polyethylene glycol diglycidyl ether (PEGDGE), and a curing reagent. They identified the complete compatibility between lignin and PEGDGE though the studies on dynamic mechanical analysis.

3.3.1.3. Ligninin rubber blends

Although the history of lignin/rubber blend started in 1949, a very little work has been performed till date [186]. The role of lignin in rubber is identified as reinforcing filler and stabilizer or antioxidant. Kumaran et al. [186] performed an extensive research on the utiliza-tion of lignins in rubber compounding and identified the improvement of many properties. They reported that the addition of lignin into rubber improved their tear, abrasion and flexural crack resistances. Košíková et al. [167] investigated the reinforcement effect of sulfur-free lignin with styrene butadiene rubber (SBR). They identified that the lignin blending with SBR influenced their vulcanizing behavior and enhanced the various physicomechanical properties significantly. In addition to that, Wang et al. [187] investigated the fabrication of latex/modified lignin blend and identified their effective water barrier properties. Processing condition of the lignin/rubber is critical for the achievement of better properties. Tibenham et al. [188] reported the hot-milling of lignin/rubber precursors with hexamethylenetetramine, which yields a vulcanizate. They also found that the modulus, tensile strength, and hardness properties were in the same order as the rubber reinforced with carbon blacks.

3.3.1.4. Lignin in polyurethane

Polyurethanes are made of diisocyanate and polyol precursors, which have been used for the highly diversified applications. Traditionally, they were made from petroleum-based synthetic polyols and nowadays soy-based polyols are also widely used as the renewable feedstock. Nakamura and his co-researchers [189] investigated the lignin-based polyurethane (PU) films using polyethylene glycol (PEG) and diphenylmethane diisocyanate (MDI). They reported the thermal behavior of new polyurethane system, which indicates that the addition of lignin to PEG enhances their T_g proportionally. The combination of lignin and PEG for the formation of polyurethane resulted in various types of microstructure such as soft and flexible and hard. Their mechanical properties were highly dependent on their distribution as well as cross-linking ability between lignin-PEG-MDI segments [190]. Sarkar et al. [191] reported the synthesis of lignin–hydroxyl terminated polybutadiene (HTPB) co-polyurethanes using toluene diisocyanate as initiator. Their characterizations showed the better properties up to 3% lignin incorporation and further increment of lignin caused the reduction in their properties [191]. Saraf et al. [192-194] made an extensive research on various aspects of lignin-based polyurethane and suggested their suitable formulations for the enhanced performance. In addition to that, various types of lignin also investigated for the fabrication of polyurethane systems [195-196]. Thring et al. [195] reported the fabrication of polyurethanes from Alcell®. They found that the increasing lignin content decreases the degree of swelling and cross-linking and causes the formation of brittle and hard structures. Yoshida et al. [196] utilized the kraft lignin for the fabrication of polyurethanes. They reported that the increasing lignin content increases the cross-link density and generally causes a hard and brittle nature. They also fabricated the polyurethane from various kraft lignins with different molecular weight and found that the cross-link density has increased with increasing molecular weight [197]. These studies conclude that the higher loadings of lignin in polyurethane caused the formation of rigid structure due to higher cross-link density and resulted in poor mechanical properties. This can be overcome by employing suitable chemistry in controlling the order of cross-linking.

3.3.2. Lignin in adhesives

Phenolic structure of lignin offers possible substitution with phenol-formaldehyde (PF) resin, which exhibits a wide range of applications as adhesives. Lignin substitute in phenol-formaldehyde (PF) formulation can vary from 30 to 50%, which exhibits similar or better performance compared to virgin PF resin. Haars et al. [198] reported the fabrication of room-temperature curing adhesives using lignin and phenoloxidases as precursor chemicals. They reported the possible use of this new bioadhesive as thermosetting glue. They also indentified the increment of water resistance during the usage in particleboard production. Mansouri et al. [199] demonstrated the fabrication of lignin adhesives without formaldehyde for wood panel. Their synthesized lignin adhesives showed better internal bond strength, which also passed required international standard specifications. They found that the newly bioadhesive from lignin exhibits many properties comparable to formaldehyde-based commercial adhesives. Schneider et al. [200] patented the new technology for the fabrication of new kind of adhesives using furfuryl alcohol and lignin employing zinc chloride-based catalyst. Lignin isolated from bagasse was also experimented for the fabrication of biobased cost effective adhesives [201-202]. The obtained adhesives were used for the purpose of particleboard and wood adhesives.

3.3.3. Lignin based carbon nanostructures

Recently, carbonaceous nanomaterials that include carbon nanotubes, carbon nanofibres, graphene/graphite nanosheets and also particulate carbon nanostructures have received an extensive importance due to their possible commercial values in diversified areas like polymeric composites, sensors, energy storage/ conversion, catalysis, filters and biology [203-208]. Traditionally, carbonaceous materials were prepared from petroleum-based precursors (liquid/gaseous hydrocarbons and carbon rich polymers such as polyacrylonitrile-PAN). As the global demand for carbon materials (nano/micro) grows continually and also the conventional sources are finite there is a need to investigate for the alternate carbon source. Thus, renewable resource-based biomaterials such as seed, oil, dried fibres as well as stem have been explored for the development of various carbon materials [209-212]. The challenge in using plant-based materials as carbon feedstock is to control the carbonizing process as well as the usage of suitable catalysis in order to achieve nanostructured materials. In addition to the larger availability of biobased feedstocks for the carbonaceous materials, it also provides eco-friendliness with the reduced carbon footprint. The biofeedstock exhibits a diversified morphology with the various combination of chemical structures, which can result in the formation of varieties of carbon nanostructures. Lignin has been widely used for the fabrication of activated carbon, however synthesizing carbon nanostructures such as particles/fibres are very new, thus next section summaries the effective uses of lignin as precursor for the fabrication of carbon nanostructures and their emerging applications [213-214].

3.3.3.1. Carbon nanoparticles

Lignin can be used as an efficient precursor in synthesizing carbonaceous nanomaterials with different morphology not only due to their carbon rich phenolic structure, but also for their capable chemical modification. Synthesizing the carbon nanoparticles with different morphology is possible by adopting various order of chemical modification as well as the processing conditions. The challenge is to inhibit the nucleation of carbon structures during the carbonization process to avoid larger particles, which normally occurs at elevated temperatures. Chemical modification can result in the formation of cross-linked structure, which normally alters the carbonization mechanism and can cause the formation of carbon nanostructures with different morphology. Babel et al. [215] reported the synthesis of KOH activated lignin-based carbon nanoparticles and their effective hydrogen storage capability. Recently, Gonugunta et al. reported the fabrication of carbon nanoparticels from lignin by adopting freeze drying process [216-217].

3.3.3.2. Lignin-based carbon nanofibres

One dimensional (1D) nanostructures such as fibrous materials receives recent attention due to their unique physicochemical properties. A wide range of fabrication techniques have been used for the fabrication of fibrous nanomaterials, among them electrospinning has been found to be an efficient technique for the fabrication of various types of fibre nanostructures using polymeric solutions as precursors [218]. The fabrication of carbon nanofibres from lignin through electrospinning has three steps and they are (i) electrospinning of lignin fibres, (ii) thermal stabilization of lignin fibres and (iii) carbonization of thermo-stabilized lignin fibres [219]. Normally, lignin exhibits poor viscoelastic properties, which creates a lot of challenges during the electrospinning process. This can be overcome by blending the lignin with other kind of synthetic polymers such poly(ethylene oxide) [220]. Figure 9 shows the electrospun lignin fibre as the precursor for the fabrication of carbon nanofibres. Dallmeyer et al. [221] investigated seven different technical lignins (isolated lignin) for the fabrication of fibrous network. None of the lignins were able to be spun into fibres without a binding polymer such as PEO. In addition to PEO, utilization of polyacrylonitrile (PAN) as binding polymer was reported by Seo et al. for the fabrication of lignin-based carbon fibres [222]. The physicochemical properties and the morphology of lignin-based carbon nanofibres can be varied by manipulating the experimental parameters. Lallave et al. [219] reported the fabrication of various types of (filled and hollow) carbon nanofibers from Alcell lignins by coaxial electrospinning. Uniqueness of their process is the successful electrospinning of lignin without binder

Figure 9. Electrospun lignin fibres (reprinted the figure with permission) [219].

polymer. Recently, Spender et al. [223] reported the rapid freezing process for the fabrication of lignin fibres with nano dimension.

3.4. Proteineous meals

As discussed in the previous sections, the major proteineous meal coproducts of biofuel industry are corn gluten meal from wet mill bioethanol and soybean meal from biodiesel industries. Different types of soy protein including soy flour (48% protein), soy protein concentrate (64% protein) and soy protein isolate (92% protein) can be extracted from soybean meal after oil extraction of soybean powder with hexane [224]. Similarly, corn gluten meal can be used to extract zein, the major protein in corn [225]. These proteins can be plasticized to produce films and formable thermoplastics. The biomaterial application of the these proteins has been investigated and reviewed extensively [225-226]. Recently, the biomaterial application of the meals themselves has attracted attentions and been studied in the form of plasticized meals as well as reinforcing fillers used in polymeric biocomposites.

3.4.1. Corn gluten meal

Corn gluten meal (CGM) is much cheaper than zein protein, thus creating more attraction compared to zein in producing thermoplastic materials. In this context, several plasticizers have been tried by many researchers for plasticization of CGM. Lawton and coworkers [227] studied the effect plasticizers such as glycerol, triethylene glycol (TEG), dibutyl tartrate, and octanoic acid on melt processing and tensile properties of CGM. In another work, di Gioia et al. [228] plasticized CGM with different plasticizers including water, glycerol, polyethylene glycols (PEG), glucose, urea, diethanolamine, and triethanolamine, at concentrations of 10–30% (dwb). They implemented dynamic mechanical thermal analysis (DMTA) to investigate the change in glass transition temperature and rheological moduli of CGM. Similarly, the effect of "polar" plasticizers (such as water, glycerol) or "amphiphilic" plasticizers (such as octanoic and palmitic acids, dibutyl tartrate and phthalate, and diacetyl tartaric acid ester of mono-diglycerides) on the glass transition temperature of the CGM/plasticizer blends have been reported [229].

Plasticized CGM has been blended with several polymers. Corradini et al. [230] blended CGM with different plastics such as starch, polyvinyl alcohol (PVA) and poly(hydroxybutyrate-co-hydroxyvalerate), PHBV, using glycerol as plasticizer. After studying the glass transition temperature of the blends, they found that these blends are immiscible in the studied compositional range. Also in terms of mechanical properties, PVA improved the flexibility while PHBV enhanced the rigidity and starch caused slight changes in mechanical properties. CGM has also been blended with poly(ε-caprolactone), PCL [231]. In this work, CGM was first plasticized using glycerol/ethanol mixture, denatured by the addition of guanidine hydrochloride (GHCl), and then blended with PCL. They used twin screw extruder and injection molding for the processing. Their results showed that chemical modification of plasticized CGM with GHCl resulted in a high percent elongation. In another work CGM was blended with poly(lactic acid), PLA, plasticized with glycerol, water and ethanol using a single screw extruder followed by compression molding [212]. Their results showed that PLA enhanced the rigidity and improved the water resistance. CGM-wood fiber biocomposites have been the point of interest in several publications. CGM in these works has been used in the form of plasticized meal. Wu et al. [232] produced pellets of CGM-wood fiber, plasticized by glycerol, water and ethanol, to manufacture injection-molded plant pots for developing low cost, biodegradable containers used in agriculture. In another study, CGM plasticized with propylene glycol was blended with a biopolymer, poly(butylene succinate) (PBS), and wood fiber to produce a biodegradable material for plastic packaging applications [233]. The CGM content varied between 10–80 wt% and it was found that the produced biomaterial exhibited relatively high tensile strength, elongation at break and water resistance as long as the CGM content was less than 30 wt%. Similarly, CGM has been plasticized with different plasticizers such as glycerol, octanoic acid, polyethylene glycol and water, and reinforced with wood fiber using a twin screw extruder [234]. The best mechanical performance was achieved when a combination of 10 wt% octanoic acid and 30 wt% water was used as plasticizer with 20 wt% wood fibre as reinforcement. The mechanical properties were improved more when the CGM matrix was blended with polypropylene, coupling agent (maleated polypropylene) and cross-linking agent (benzoyl peroxide) with 50 wt% wood fibre [234].

3.4.2. Soybean meal

Soybean meal (SM) is still finding its way within scientific researches towards biobased material (biomaterial) applications. The number of publications on this topic is limited compared to corn gluten meal. SM has been characterized for its chemical composition, moisture content, thermal behavior and infrared spectrum and its potential as a particulate filler in value-added biocomposites was evaluated by compounding with polycaprolactone (PCL) [235]. The composite of PCL/SM (70/30) was prepared by extrusion and injection molding and then tested for mechanical properties such as tensile, flexural and impact. It has been observed that PCL/SM composite exhibited higher tensile/flexural modulus, but lower strength, elongation and impact strength compared to PCL. At the same time, the resulted biocomposite had relatively less cost than PCL itself. Thus, addition of SM to PCL increases the rigidity, but the particle-matrix adhesion needs to be improved.

SM has been plasticized and blended with other polymers such as polycaprolactone (PCL), poly(butylenes succinate) (PBS), poly(butylene adipate terephthalate) (PBAT) [236-237], and natural rubber [238-239]. SM was plasticized and destructurized successfully using glycerol and urea in a twin screw extruder and then blended with biodegradable polyesters, PCL and PBS [236]. As a result of destructurization phenomenon, improvement occurred in mechanical properties of the protein-based blends. In another work, SM was plasticized using glycerol in presence of two different denaturants (destructurizers), and the resulted thermoplastic SM was blended with different polyesters, PBS, PCL and PBAT [237]. Taguchi experimental design was adopted to investigate the effect of each constituent on the tensile properties of the final blend. Wu et al. [238] produced vulcanized blend of natural rubber and 50 wt% SM. The rubber phase was embedded by the SM matrix suggesting the interaction between phases, also approved by the increase in the glass transition temperature of the rubber phase. The produced blend exhibited good elasticity and water resistance.

Another area of research conducted on SM is producing edible films from defatted SM for food packaging applications [240-241]. For this purpose, SM was fermented in a soybean meal solution (15 g/100 ml of water) by inoculation with Bacillus subtilis bacteria fermentated under optimum conditions of 33°C and pH 7.0-7.5 for 33 h. Then, the fermented soybean solution was heated for 20 min at 75°C with 2-3 ml glycerol added to the solution to overcome film brittleness. The filtered solution was finally casted in a petri dish to produce films. Increasing amount of plasticizer in the fermented film led to a decrease in tensile strength and an increase in % elongation of the film compared to the ordinary soybean film. Moreover, the SM-based film exhibited higher water vapor permeability. On the other hand, experiments showed that growth inhibition of the produced SM-based film in the agar media containing E. coli was much higher than the ordinary soy protein film. These results indicated that the fermented SM-based films can be used as a new packaging material to extend the shelf-life of foods; however mechanical and physical properties need to be improved for more industrial applications.

3.5. Crude glycerol

Value-added uses of crude glycerol from biodesiel industries are highly diversified and it can be classified in to following categories, (i) chemicals/monomers: extraction and chemical/ biological conversion approach for the synthesis of various precursors, (ii) plasticizer: use as low cost plasticizing agent in various biomaterials applications, (iii) hydrogen generation: as the source for the generation of green hydrogen for energy applications, (iv) carbon source: carbon sole source for bacterial growth, which effectively used for the generation of bioplastics and (v) polyesters: combine with suitable organic acids and forms polyesters for various materials applications. In materials point of view, crude glycerol's application in plasticising as well as polyesters formation receives an immense attention and is discussed in this section.

As the global demand increases for the development of biodegradable materials, starch-based plastics receives an immense attention. The challenge is their unusual inter and intra molecular hydrogen bonds, which reduces their plastic performance [242]. This can be improved by incorporating plasticizers and manipulating processing conditions. Various plasticizers have been used on thermoplastic starch such as glycerol [243], glycol [244], sorbitol [245] and sugar [246]. Among them glycerol has been extensively used for the plasticization of starch-based materials. In addition to that, glycerol has also been used as an effective plasticizer in various other natural materials such as cellulose [247], chitosan [248], gelatin [249], DDG [250] and protein [233]. When these materials are plasticised with glycerol, their elasticity and toughness have been increased significantly with decreased brittleness. Numerous literatures are available for the glycerol plasticized natural products and most of the researchers have practiced pure glycerol as the precursor [233, 247-251]. Increasing trend of crude glycerol as plasticizer can be found in the literature for the development of various types of biomaterials [252-259].

Glycerol can be used to create polymeric materials by exploring their polyfunctional reactivity and the obtained polymers can have various applications as polyols substitutes, lubricant, raw materials to produce resins, polyesters and polyurethanes [260]. Among them, polyesters and polyurethanes receive more importance for their large volume biomaterial applications. The glycerol can easily react with carboxylic organic acids and forms polyesters, which can be used for various materials applications [260-261]. Carnahan et al. [262] demonstrated the synthesis of glycerol-based polyesters using succinic and adipic acids. Likewise, Tang et al. [263] synthesized the aliphatic polyesters from glycerol using sebacic acid with elastic properties. In both the cases they were using pure glycerol as a precursor and more publications are also available for fabrication of various other types of polyesters [264-265]. However, very few research works have been published based on the crude glycerol as feedstock for polyester synthesis. Brioude et al. [261] reported the fabrication of new polyester from crude glycerol and adipic acid through the bulk polymerization. They found that the newly developed crude glycerol-based polyester was amorphous in nature and has good mechanical and thermal stability. Similarly de Moura et al. [260] reported the polymerization between the glycerol and mono/bi-functional organic groups for the fabrication of new class of polyesters. They were able to synthesise two different classes of polymers with variable thermal stability. Their ultimate aim was to utilize these newly developed polyesters as matrix for the fabrication of natural fibre-reinforced composite materials. The obtained features (physicochemical) of these newly synthesised polyesters can also be used as modifiers for various types of thermosets (epoxy resins) as well as polyurethanes. Urea-formaldehyde (UF) has been widely known as an excellent adhesive used in wood panel fabrication. The challenge is to overcome their limited moisture resistance. The polyols made form glycerol can be used to enhance their moisture resistance properties. Recently, fungi-based biological transformation of crude glycerol in to various value-added products receives significant attention. Fungus exhibits more tolerance against the various impurities that exist in crude glycerol and found to be a suitable candidate for the biotransformations. Nicol et al. [266] reviewed the various research work on effective bioconversion of crude glycerol into many value added chemicals/materials employing fungus. They conclude that these techniques need to be further investigated for the extended applications. It is visible that glycerol will become as a high potential feedstock for various chemicals and tends to replace various existing petroleum derived products [267].

4. Conclusions

Growing energy demand for transportation, necessity of reducing the dependency on fossil fuel and the emerging environmental concerns about greenhouse gas emission have enhanced the growth of biofuel industries enormously and created new policy mandates by many countries. This caused the generation of a huge amount of under-valued coproduces of different types including distillers' grains, bagasse, lignin, protein-rich meals and crude glycerol. In order to claim biofuel production as the sustainable technology of future, it is necessary to create value-addition to these coproducts. This also provides solution to the emerging issues on the environmental impact of the accumulation of these coproducts. Moreover, the technological development related to the value-added uses of biofuel copro-ducts in biomaterials applications, is expected to (i) create new ecofriendly products, (ii) strengthen the bioeconomy and (iii) generate more rural job opportunity.

Among these coproducts, dried distillers' grains with solubles (DDGS), from dry mill ethanol industry, and the protein-rich meals such as corn gluten meal (CGM) and soybean meal (SM), respectively from wet mill ethanol and biodiesel industries, are traditionally used for animal feed applications. However, recent researches show the huge potential of these coproducts for biomaterial applications. DDGS can be used as filler or even as reinforcing phase in polymeric biocomposites by integrating pre-treatment, compatibilization and proper processing techni-ques. Moreover, it can be converted into a thermoplastic by chemical modifications such as acetylation and cyanoethylation. Also, it can be used for producing bioadhesive or extraction of fiber, oil and protein to be used for biomaterial usages.

Due to their high content of protein, CGM and SM can be plasticized using several plasticizers. Such plasticized meals have been blended successfully with several thermoplastics or been reinforced with wood fiber to create biomaterials with balanced properties. Besides, fermen-tation of these proteineous coproducts can provide new opportunities for the development of biobased films for food packaging applications.

It has been a long time since bagasse, the sugarcane ethanol coproduct, has been used in thermoset-fiber composites for particleboard production. In addition to that, bagasse is recently considered as the reinforcing filler compounded with different thermoplastics. Also, it has been shown that bagasse can turn into a thermo-formable material through esterification.

Lignin, the major coproduct of lignocellulosic ethanol, is being upgraded from its traditional utilization, i.e. burning, into various value-added commercial products such as polymer blends/composites, adhesives, and carbon fibres.

The industrial applications of pure glycerol in different sections are well established. Consequently, crude glycerol coproduced in biodiesel industry has huge opportunities in biomaterial applications including chemicals/monomers, plasticizer, hydrogen generation, carbon source for bacterial growth and polyesters production.

Novel innovative research in utilizing biofuel coproducts for biomaterials applications (biopolymers and biocomposites) impends the value-added high-end uses in the automotive, packaging and other structural/durable sectors. This is expected to create a fundamental change in materials point of view by utilizing renewable resources as feedstocks. In addition to that, these biofuel coproducts also create opportunity for the fabrication of various nano-structured materials for the high value applications.

Acknowledgements

Authors are thankful to the, Ontario Research Fund, Research Excellence, Round-4 (ORF RE04) from the Ontario Ministry of Economic Development and Innovation (MEDI); Ontario Ministry of Agriculture, Food, and Rural Affairs' (OMAFRA) New Directions and Alternative Renewable Fuels "Plus" Research Programs; OMAFRA-University of Guelph Bioeconomy-Industrial Uses Program; the Natural Sciences and Engineering Research Council of Canada (NSERC) Discovery grant individuals to M. Misra and A. K. Mohanty; NSERC Biomaterials and Chemicals Strategic Research Network; AUTO21-NCE (Canada's automotive R&D program); the Hannam Soybean Utilization Fund (HSUF) and the Grain Farmers of Ontario (GFO) fund for supporting research on various aspects of biobased materials, biofuel co-products and bionanotechnology at the University of Guelph's Bioproducts Discovery and Development Centre. N. Zarrinbakhsh acknowledges highly qualified personnel (HQP) scholarship from OMAFRA 2009.

Author details

S. Vivekanandhan[1,2], N. Zarrinbakhsh[1,2], M. Misra[1,2] and A. K. Mohanty[1,2]

1 Bioproducts Discovery and Development Centre, Department of Plant Agriculture, Crop Science Building, University of Guelph, Guelph, ON, N1G 2W1, Canada

2 School of Engineering, Thornbrough Building, University of Guelph, Guelph, ON, N1G 2W1, Canada

References

[1] 2012 The Outlook for Energy: A View to 2040 – Highlights. http://www.exxonmobil. com /corporate/files/news_pub_eo2012_highlights.pdf (Accessed on 30th July 2012).

[2] Weischer L, Wood D, Ballesteros A, Fu-Bertaux X. Grounding Green Power-Bottom-up Perspectives on Smart Renewable Energy Policy in Developing Countries, Climate & Energy Paper Series, The German Marshall Fund of the United States, Washington, 2011.

[3] Pereira MG, Camacho CF, Freitas MAV, da Silva NF. The Renewable Energy Market in Brazil: Current Status and Potential. Renewable and Sustainable Energy Reviews 2012;16 3786-3802.

[4] Rosegrant MW, Zhu TJ, Msangi S, Sulser T. Global Scenarios for Biofuels: Impacts and Implications. Applied Economic Perspectives and Policy 2008;30 495-505.

[5] Biodiesel Benefits, http://www.Biodiesel.com/index.php/biodiesel/biodiesel_bene-fits_why_use_biodiesel/. (accessed on 30th July 2012).

[6] Rajagopal D, Sexton SE, Roland-Holst D, Zilberman D. Challenge of Biofuel: Filling the Tank without Emptying the Stomach? Environmental Research Letters 2007;2 044004 (doi:10.1088/1748-9326/2/4/044004).

[7] Ragauskas AJ, Williams CK, Davison BH, Britovsek G, Cairney J, Eckert CA, et al. The Path Forward for Biofuels and Biomaterials. Science 2006;311 484-489.

[8] Walker GM. Bioethanol: Science and Technology of Fuel Alcohol, Graeme M. Wakler & Ventus Publishing ApS, Avilable online at http://bookboon.com/, 2010.

[9] Bhullar AS, Deo B, Sachdeva J. Transport Biofuel Production, Trade-Offs, and Promotion Policies in Canada—a Review. Agricultural Economics Research Review 2012;25 137-150.

[10] Letete T, Blottnitz H. Biofuel Policy in South Africa: A Critical Analysis, In: Rainer J, Dominik R, Bioenergy for Sustainable Development in Africa, Springer, 2012 191-199.

[11] Khanna M, Önal H, Crago CL, Mino K. Can India Meet Biofuel Policy Targets? Implications for Food and Fuel Prices. American Journal of Agricultural Economics 2012 doi: 10.1093/ajae/aas040.

[12] Wang Q, Tian Z. Biofuels and the Policy Implications for China. Asian Pacific Economic Literature 2011;25 161-168.

[13] Abdullah AZ, Salamatinia B, Mootabadi H, Bhatia S. Current Status and Policies on Biodiesel Industry in Malaysia as the World's Leading Producer of Palm Oil. Energy Policy 2009;37 5440-5448.

[14] Sparks GD, Ortmann GF. Global Biofuel Policies: A Review. Agrekon 2011;50 59-82.

[15] Lynd LR, Laser MS, Bransby D, Dale BE, Davison B, Hamilton R, et al. How Biotech Can Transform Biofuels. Nature Biotechnology 2008;26 169-172.

[16] Paulson ND, Ginder RG. The Growth and Direction of the Biodiesel Industry in the United States: Center for Agricultural and Rural Development, Iowa State University; 2007.

[17] Fernando S, Adhikari S, Chandrapal C, Murali N. Biorefineries: Current Status, Challenges, and Future Direction. Energy & Fuels 2006;20 1727-1737.

[18] OECD-FAO Agricultural Outlook 2011, Organisation for Economic Co-operation and Development; 2011. (DOI: 10.1787/agr_outlook-2011-en).

[19] OECD-FAO Agricultural Outlook 2011, Development of the World Ethanol Market. (http://dx.doi.org/10.1787/888932426467). Organisation for Economic Co-operation and Development; 2011.

[20] OECD-FAO Agricultural Outlook 2011, Evolution of Global Ethanol Production by Feedstocks Used (http://dx.doi.org/10.1787/888932426562). Organisation for Economic Co-operation and Development; 2011.

[21] OECD-FAO Agricultural Outlook 2011, Development of the World Biodiesel Market (http://dx.doi.org/10.1787/888932426486), Organisation for Economic Co-operation and Development; 2011.

[22] OECD-FAO Agricultural Outlook 2011, Evolution of Global Biodiesel Production by Feedstocks Used (http://dx.doi.org/10.1787/888932426581), Organisation for Economic Co-operation and Development; 2011.

[23] Rosentrater KA. Ethanol processing co-products: economics, impacts, sustainability. In: Eaglesham A, Hardy RWF, editors. Agricultural biofuels: technology, sustainability and profitability. Proceedings of the National Agricultural Biotechnology Council's 19th Annual Conference; 2007 May 22–24; Brookings (SD), USA. Ithaca (NY): National Agricultural Biotechnology Council; 2007. p. 105-26.

[24] RFA Renewable Fuels Association. Industry resources: co-products; Available from: http://www.ethanolrfa.org/pages/industry-resources-coproducts (Accessed on 1st October 2012).

[25] Yu P, Nuez-Ortin WG. Relationship of Protein Molecular Structure to Metabolisable Proteins in Different Types of Dried Distillers Grains with Solubles: A Novel Approach. British Journal of Nutrition 2010;104 1429-37.

[26] Kim Y, Mosier NS, Hendrickson R, Ezeji T, Blaschek H, Dien B, et al. Composition of Corn Dry-Grind Ethanol by-Products: DDGS, Wet Cake, and Thin Stillage. Bioresource Technology 2008;99 5165-5176.

[27] Dong FM, Rasco BA. The Neutral Detergent Fiber, Acid Detergent Fiber, Crude Fiber, and Lignin Contents of Distillers' Dried Grains with Solubles. Journal of Food Science 1987;52 403.

[28] Corn Refiners A. Corn Wet Milled Feed Products: http://www.corn.org/wp-content/uploads/2009/12/Feed2006.pdf; 2006. (Accessed on 17th September 2012)

[29] Getachew G, Robinson PH, DePeters EJ, Taylor SJ. Relationships between Chemical Composition, Dry Matter Degradation and in Vitro Gas Production of Several Ruminant Feeds. Animal Feed Science and Technology 2004;111 57-71.

[30] Bell JM, Keith MO. A Survey of Variation in the Chemical Composition of Commercial Canola Meal Produced in Western Canadian Crushing Plants. Canadian Journal of Animal Science 1991;71 469-480.

[31] Makkar HPS, Aderibigbe AO, Becker K. Comparative Evaluation of Non-Toxic and Toxic Varieties of Jatropha Curcas for Chemical Composition, Digestibility, Protein Degradability and Toxic Factors. Food Chemistry 1998;62 207-215.

[32] Anderson JL, Schingoethe DJ, Kalscheur KF, Hippen AR. Evaluation of Dried and Wet Distillers Grains Included at Two Concentrations in the Diets of Lactating Dairy Cows. Journal of Dairy Science 2006;89 3133-3142.

[33] Gill RK, VanOverbeke DL, Depenbusch B, Drouillard JS, DiCostanzo A. Impact of Beef Cattle Diets Containing Corn or Sorghum Distillers Grains on Beef Color, Fatty Acid Profiles, and Sensory Attributes. Journal of Animal Science 2008;86 923-935.

[34] Wahlstrom RC, German CS, Libal GW. Corn Distillers Dried Grains with Solubles in Growing-Finishing Swine Rations. Journal of Animal Science 1970;30 532-535.

[35] Oryschak M, Korver D, Zuidhof M, Meng X, Beltranena E. Comparative Feeding Value of Extruded and Nonextruded Wheat and Corn Distillers Dried Grains with Solubles for Broilers. Poultry Science 2010;89 2183-2196.

[36] Loar II RE, Schilling MW, McDaniel CD, Coufal CD, Rogers SF, Karges K, et al. Effect of Dietary Inclusion Level of Distillers Dried Grains with Solubles on Layer Performance, Egg Characteristics, and Consumer Acceptability. The Journal of Applied Poultry Research 2010;19 30-37.

[37] Cozannet P, Lessire M, Gady C, Métayer JP, Primot Y, Skiba F, et al. Energy Value of Wheat Dried Distillers Grains with Solubles in Roosters, Broilers, Layers, and Turkeys. Poultry Science 2010;89 2230-2241.

[38] Whitney TR, Lupton CJ. Evaluating Percentage of Roughage in Lamb Finishing Diets Containing 40% Dried Distillers Grains: Growth, Serum Urea Nitrogen, Nonesterified Fatty Acids, and Insulin Growth Factor-1 Concentrations and Wool, Carcass, and Fatty Acid Characteristics. Journal of Animal Science 2010;88 3030-3040.

[39] Webster CD, Tidwell JH, Yancey DH. Evaluation of Distillers' Grains with Solubles as a Protein Source in Diets for Channel Catfish. Aquaculture 1991;96 179-190.

[40] Shelby RA, Lim C, Yildrim-Aksoy M, Klesius PH. Effect of Distillers Dried Grains with Solubles incorporated Diets on Growth, Immune Function and Disease Resistance in Nile Tilapia (Oreochromis Niloticus L.). Aquaculture Research 2008;39 1351-1353.

[41] Ayadi FY, Muthukumarappan K, Rosentrater KA, Brown ML. Twin-Screw Extrusion Processing of Distillers Dried Grains with Solubles (DDGS)-Based Rainbow Trout (Oncorhynchus Mykiss) Feeds, ASABE Meeting Presentation, Pennsylvania, Paper Number: 1008337 2010.

[42] Tidwell JH, Webster CD, Clark JA, D'Abramo LR. Evaluation of Distillers Dried Grains with Solubles as an Ingredient in Diets for Pond Culture of the Freshwater Prawn Macrobrachium Rosenbergii. Journal of the World Aquaculture Society 1993;24 66-70.

[43] Babcock BA, Hayes DJ, Lawrence JD. Using Distillers Grains in the U.S. And International Livestock and Poultry Industries. USA: Midwest Agribusiness Trade Research and Information Center; 2008.

[44] Hoffman LA, Baker A. Market issues and prospects for U.S. distillers' grains supply, use, and price relationships. Washington, DC: United States Department of Agriculture, Economic Research Service; 2010 Dec. Outlook No.: FDS-10k-01.

[45] A Guide to Distiller's Dried Grains with Solubles (DDGS), U.S. Grains Council, http://www.ethanolrfa.org/page/-/rfa-association-site/studies/2012_DDGS_Handbook.pdf?nocdn=1 (Accessed on 2nd December 2012).

[46] Tokgoz S, Elobeid A, Fabiosa J, Hayes DJ, Babcock BA, Yu T-H, et al. Emerging Biofuels: Outlook of Effects on U.S. Grain, Oilseed, and Livestock Markets. Iowa State University: Center for Agricultural and Rural Development; 2007.

[47] Canadian Food Inspection Agency. Rg-6 Regulatory Guidance: Ethanol Distillers' Grains for Livestock Feed. Available Online at: http://www.inspection.gc.ca/animals/feeds/regulatory-guidance/rg-6/eng/1329275341920/1329275491608. (Accessed on 30th July 2012).

[48] AllAboutFeed (2011) US DDGS exports reach record high in 2010. http://www.allaboutfeed.net/news/us-ddgs-exports-reach-record-high-in-2010-5303.html. (Accessed on February 2012).

[49] Chevanan N, Muthukumarappan K, Rosentrater KA. Extrusion Studies of Aquaculture Feed Using Distillers Dried Grains with Solubles and Whey. Food and Bioprocess Technology 2009;2 177-185.

[50] Rasco BA, Rubenthaler G, Borhan M, Dong FM. Baking Properties of Bread and Cookies Incorporating Distillers' or Brewer's Grain from Wheat or Barley. Journal of Food Science 1990;55 424-429.

[51] Morey RV, Hatfield DL, Sears R, Tiffany DG. Characterization of Feed Streams and Emissions from Biomass Gasification/Combustion at Fuel Ethanol Plants. ASABE Meeting Presentation, Oregon, Paper Number: 064180 2006.

[52] Morey RV, Tiffany DG, Hatfield DL. Biomass for Electricity and Process Heat at Ethanol Plants. Applied Engineering in Agriculture 2006;22 723.

[53] Xu W, Reddy N, Yang Y. An Acidic Method of Zein Extraction from DDGS. Journal of Agricultural and Food Chemistry 2007;55 6279-6284.

[54] Xu W, Reddy N, Yang Y. Extraction, Characterization and Potential Applications of Cellulose in Corn Kernels and Distillers' Dried Grains with Solubles (DDGS). Carbohydrate Polymers 2009;76 521-527.

[55] Singh N, Cheryan M. Extraction of Oil from Corn Distillers Dried Grains with Solubles. Transactions of the ASAE 1998;41 1775-1777.

[56] Bruinsma K, Endres DL. Solvent Extraction of Oil from Distillers Dried Grains and Methods of Using Extraction Products. US Patent 2012; US 8,227,015 B2.

[57] Weber AD. The Relative Value of Cottonseed Meal, Linseed Meal, and Corn Gluten Meal in Fattening Cattle Rations. Journal of Animal Science 1934;1934 70-72.

[58] Umunna NN, Klopfenstein TJ, Hasimoglu S, Woods WR. Evaluation of Corn Gluten Meal with Urea as a Source of Supplementary Nitrogen for Growing Calves and Lambs. Animal Feed Science and Technology 1982;7 375-385.

[59] Fellner V, Belyea RL. Maximizing Gluten Feed in Corn Silage Diets for Dairy Cows. Journal of Dairy Science 1991;74 996-1005.

[60] Babidis V, Florou-Paneri P, Kufidis D, Christaki E, Spais AB, Vassilopoulos V. The Use of Corn Gluten Meal Instead of Herring and Meat Meal in Broiler Diets and Its Effect on Performance, Carcass Fatty Acid Composition and Othercarcass Characteristics. Archiv für Geflügelkunde 2002;66 145-150.

[61] Shurson G, Spiehs M, Whitney M. The Use of Maize Distiller's Dried Grains with Solubles in Pig Diets. Pig News and Information 2004;25 75N-83N.

[62] Funaba M, Oka Y, Kobayashi S, Kaneko M, Yamamoto H, Namikawa K, et al. Evaluation of Meat Meal, Chicken Meal, and Corn Gluten Meal as Dietary Sources of Protein in Dry Cat Food. Canadian Journal of Veterinary Research 2005;69 299-304.

[63] Wu YV, Rosati RR, Sessa DJ, Brown PB. Evaluation of Corn Gluten Meal as a Protein Source in Tilapia Diets. Journal of Agricultural and Food Chemistry 1995;43 1585-1588.

[64] Christians NE. Preemergence Weed Control Using Corn Gluten Meal. US Patent 1994; RE34594.

[65] Biopesticides registration action document, Corn Gluten Meal (PC Code 100137), US Environmental Protection Agency, Office of Pesticide Programs, Available online at: http://www.epa.gov/pesticides/chem_search/reg_actions/registration/decision_PC-100137_4-Mar-03.pdf. (Accessed: December 2012).

[66] Anselmo Filho P, Badr O. Biomass Resources for Energy in North-Eastern Brazil. Applied Energy 2004;77 51-67.

[67] Pandey A, Soccol CR, Nigam P, Soccol VT. Biotechnological Potential of Agro-Industrial Residues. I: Sugarcane Bagasse. Bioresource Technology 2000;74 69-80.

[68] Srinivasan VR, Han YW. Utilization of Bagasse. In: Hajny GJ, Reese ET, editors. Cellulases and Their Applications: American Chemical Society; 1969, 447-460.

[69] Mbohwa C, Fukuda S. Electricity from Bagasse in Zimbabwe. Biomass and Bioenergy 2003;25 197-207.

[70] van den Broek R, van den Burg T, van Wijk A, Turkenburg W. Electricity Generation from Eucalyptus and Bagasse by Sugar Mills in Nicaragua: A Comparison with Fuel Oil Electricity Generation on the Basis of Costs, Macro-Economic Impacts and Environmental Emissions. Biomass and Bioenergy 2000;19 311-335.

[71] Scaramucci JA, Perin C, Pulino P, Bordoni OFJG, da Cunha MP, Cortez LAB. Energy from Sugarcane Bagasse under Electricity Rationing in Brazil: A Computable General Equilibrium Model. Energy Policy 2006;34 986-992.

[72] Krüger H, Berndt W, Schwartzkopff U, Reitter FJ, Höpner T, Mühlig H-J. Production of Paper Pulp from Sugar Mill Bagasse. US Patent 1981; US 4260452.

[73] Shaikh AJ. Blending of Cotton Stalk Pulp with Bagasse Pulp for Paper Making. Biological Wastes 1990;31 37-43.

[74] Widyorini R, Xu J, Umemura K, Kawai S. Manufacture and Properties of Binderless Particleboard from Bagasse I: Effects of Raw Material Type, Storage Methods, and Manufacturing Process. Journal of Wood Science 2005;51 648-654.

[75] Zayed G, Abdel-Motaal H. Bio-Production of Compost with Low Ph and High Soluble Phosphorus from Sugar Cane Bagasse Enriched with Rock Phosphate. World Journal of Microbiology and Biotechnology 2005;21 747-752.

[76] Nigam P. Investigation of Some Factors Important for Solid-State Fermentation of Sugar Cane Bagasse for Animal Feed Production. Enzyme and Microbial Technology 1990;12 808-811.

[77] De Filippis P, Borgianni C, Paolucci M, Pochetti F. Gasification Process of Cuban Bagasse in a Two-Stage Reactor. Biomass and Bioenergy 2004;27 247-252.

[78] Laser M, Schulman D, Allen SG, Lichwa J, Antal Jr MJ, Lynd LR. A Comparison of Liquid Hot Water and Steam Pretreatments of Sugar Cane Bagasse for Bioconversion to Ethanol. Bioresource Technology 2002;81 33-44.

[79] Cuzens JC, Miller JR. Acid Hydrolysis of Bagasse for Ethanol Production. World Renewable Energy Congress IV Renewable Energy, Energy Efficiency and the Environment 1997;10 285-290.

[80] Teixeira LC, Linden JC, Schroeder HA. Optimizing Peracetic Acid Pretreatment Conditions for Improved Simultaneous Saccharification and Co-Fermentation (SSCF) of Sugar Cane Bagasse to Ethanol Fuel. Renewable Energy Energy Efficiency, Policy and the Environment 1999;16 1070-1073.

[81] Kim S, Dale BE. Global Potential Bioethanol Production from Wasted Crops and Crop Residues. Biomass and Bioenergy 2004;26 361-375.

[82] SugarCane.org. Brazilian Sugarcane Harvest. http://sugarcane.org/media-center/sugarcane-statistics/brazilian-sugarcane-harvest. (Accessed on 2nd December 2012).

[83] Taherzadeh MJ, Karimi K. Acid-Based Hydrolysis Processes for Ethanol from Lignocellulosic Materials: A Review. BioResources 2007; 2 472-499.

[84] Taherzadeh MJ, Karimi K. Enzymatic-Based Hydrolysis Processes for Ethanol from Lignocellulosic Materials: A Review. BioResources 2007; 2 707-738.

[85] Kumar S, Mohanty AK, Erickson L, Misra M. Lignin and Its Applications with Polymers. Journal of Biobased Materials and Bioenergy 2009;3 1-24.

[86] Mohanty AK, Misra M, Drzal LT. Sustainable Bio-Composites from Renewable Resources: Opportunities and Challenges in the Green Materials World. Journal of Polymers and the Environment 2002;10 19-26.

[87] Calvo-Flores FG, Dobado JA. Lignin as Renewable Raw Material. ChemSusChem 2010;3 1227-1235.

[88] Vishtal AG, Kraslawski A. Challenges in Industrial Applications of Technical Lignins. BioResources 2011;6 3547-3568.

[89] Stewart D. Lignin as a Base Material for Materials Applications: Chemistry, Application and Economics. Industrial Crops and Products 2008;27 202-207.

[90] Naik SN, Goud VV, Rout PK, Dalai AK. Production of First and Second Generation Biofuels: A Comprehensive Review. Renewable and Sustainable Energy Reviews 2010;14 578-597.

[91] Adler E. Lignin Chemistry—Past, Present and Future. Wood Science and Technology 1977;11 169-218.

[92] Sahoo S, Misra M, Mohanty AK. Enhanced Properties of Lignin-Based Biodegradable Polymer Composites Using Injection Moulding Process. Composites Part A: Applied Science and Manufacturing 2011, 42 1710–1718.

[93] Sahoo S, Misra M, Mohanty AK. Effect of Compatibilizer and Fillers on the Properties of Injection Molded Lignin Based Hybrid Green Composites. Journal of Applied Polymer Science 2012. DOI: 10.1002/APP.37667

[94] Hu TQ. Chemical Modification, Properties, and Usage of Lignin: Springer; 2002.

[95] Suhas, Carrott PJM, Ribeiro Carrott MML. Lignin–from Natural Adsorbent to Activated Carbon: A Review. Bioresource Technology 2007;98 2301-2312.

[96] Reisch MS. Getting the Steel Out. Chemical & Engineering News, 2011;89 10-14.

[97] United States Department of Agriculture, Economic Research Service. World Protein Meal Production. 2012. http://www.ers.usda.gov/dataproducts/oil-crops-yearbook.aspx. (Accesed on 1st December 2012).

[98] Faostat. http://faostat.fao.Org. (Accesed on 1st December 2012).

[99] United States National Biodiesel Board. US Biodiesel Production, http://www.biodiesel.org/production/production-statistics. (Accesed on 1st December 2012).

[100] Lang X, Dalai AK, Bakhshi NN, Reaney MJ, Hertz PB. Preparation and Characterization of Bio-Diesels from Various Bio-Oils. Bioresource Technology 2001;80 53-62.

[101] Sarin R, Sharma M, Sinharay S, Malhotra RK. Jatropha–Palm Biodiesel Blends: An Optimum Mix for Asia. Fuel 2007;86 1365-1371.

[102] Meher LC, Kulkarni MG, Dalai AK, Naik SN. Transesterification of Karanja (Pongamia Pinnata) Oil by Solid Basic Catalysts. European Journal of Lipid Science and Technology 2006;108 389-397.

[103] Pramanik K. Properties and Use of Jatropha Curcas Oil and Diesel Fuel Blends in Compression Ignition Engine. Renewable Energy 2003;28 239-248.

[104] Gerpen JV. Biodiesel Processing and Production. Fuel Processing Technology 2005;86 1097-1107.

[105] Thompson JC, He BB. Characterization of Crude Glycerol from Biodiesel Production from Multiple Feedstocks. Applied Engineering in Agriculture 2006;22 261-265.

[106] Meher LC, Vidya Sagar D, Naik SN. Technical Aspects of Biodiesel Production by Transesterification—a Review. Renewable and Sustainable Energy Reviews 2006;10 248-268.

[107] Pachauri N, He B. Value-Added Utilization of Crude Glycerol from Biodiesel Production: A Survey of Current Research Activities. ASABE Meeting Presentation, Oregon, Paper Number: 066223 2006.

[108] Mu Y, Teng H, Zhang D-J, Wang W, Xiu Z-L. Microbial Production of 1, 3-Propane-
diol by Klebsiella Pneumoniae Using Crude Glycerol from Biodiesel Preparations. Biotechnology Letters 2006;28 1755-1759.

[109] Soares RR, Simonetti DA, Dumesic JA. Glycerol as a Source for Fuels and Chemicals
by Low Temperature Catalytic Processing. Angewandte Chemie 2006;118 4086-4089.

[110] Mothes G, Schnorpfeil C, Ackermann J-U. Production of PHB from Crude Glycerol.
Engineering in Life Sciences 2007;7 475-479.

[111] Zhou C-HC, Beltramini JN, Fan Y-X, Lu GQM. Chemoselective Catalytic Conversion
of Glycerol as a Biorenewable Source to Valuable Commodity Chemicals. Chemical
Society Reviews, 2008;37 527-549.

[112] Cockerill S, Martin C. Are Biofuels Sustainable? The Eu Perspective. Biotechnology
for Biofuels 2008;1. doi:10.1186/1754-6834-1-9.

[113] Bergeron C, Carrier DJ, Ramaswamy S. Biorefinery Co-Products: Phytochemicals,
Primary Metabolites and Value-Added Biomass Processing: Wiley; 2012.

[114] Solomon BD. Biofuels and Sustainability. Annals of the New York Academy of Sciences 2010;1185 119-134.

[115] Biofuels and Sustainability Issues, http://www.biofuelstp.eu/sustainability.html (Accessed on 26th July 2012).

[116] Cook JH, Beyea J, Keeler KH. Potential Impacts of Biomass Production in the United
States on Biological Diversity. Annual Review of Energy and the Environment
1991;16 401-431.

[117] Giampietro M, Ulgiati S, Pimentel D. Feasibility of Large-Scale Biofuel Production.
BioScience 1997;47 587-600.

[118] Julson JL, Subbarao G, Stokke DD, Gieselman HH, Muthukumarappan K. Mechanical Properties of Biorenewable Fiber/Plastic Composites. Journal of Applied Polymer
Science 2004;93 2484-2493.

[119] Tisserat BH, Reifschneider L, O'Kuru RH, Finkenstadt VL. Mechanical and Thermal
Properties of High Density Polyethylene–Dried Distillers Grains with Solubles Composites. BioResources 2012;8 59-75.

[120] Wu Q, Mohanty AK. Renewable Resource Based Biocomposites from Coproduct of
Dry Milling Corn Ethanol Industry and Castor Oil Based Biopolyurethanes. Journal
of Biobased Materials and Bioenergy 2007;1 257-265.

[121] Tatara RA, Suraparaju S, Rosentrater KA. Compression Molding of Phenolic Resin
and Corn-Based DDGS Blends. Journal of Polymers and the Environment 2007;15
89-95.

[122] Tatara RA, Rosentrater KA, Suraparaju S. Design Properties for Molded, Corn-Based DDGS-Filled Phenolic Resin. Industrial Crops and Products 2009;29 9-15.

[123] Cheesbrough V, Rosentrater KA, Visser J. Properties of Distillers Grains Composites: A Preliminary Investigation. Journal of Polymers and the Environment 2008;16 40-50.

[124] Li Y, Susan Sun X. Mechanical and Thermal Properties of Biocomposites from Poly(Lactic Acid) and DDGS. Journal of Applied Polymer Science 2011;121 589-597.

[125] Zarrinbakhsh N, Mohanty AK, Misra M. Fundamental Studies on Water-Washing of the Corn Ethanol Coproduct (DDGS) and Its Characterization for Biocomposite Applications. Biomass and Bioenergy 2013; http://dx.doi.org/10.1016/j.biombioe.2013.02.016.

[126] Zarrinbakhsh N, Misra M, Mohanty AK. Biodegradable Green Composites from Distiller's Dried Grains with Solubles (DDGS) and Polyhydroxy(Butyrate-Co-Valerate) (PHBV)-Based Bioplastic. Macromolecular Materials and Engineering 2011;296 1035-1045.

[127] Muniyasamy S, Reddy MM, Misra M, Mohanty AK. Biodegradable Green Composites from Bioethanol Co-Product and Poly(Butylene Adipate-Co-Terephatalate). Industrial Crops and Products 2013;43 812– 819.

[128] Saunders JA, Rosentrater KA. Properties of Solvent Extracted Low-Oil Corn Distillers Dried Grains with Solubles. Biomass and Bioenergy 2009;33 1486-1490.

[129] Srinivasan R, Singh V, Belyea RL, Rausch KD, Moreau RA, Tumbleson ME. Economics of Fiber Separation from Distillers Dried Grains with Solubles (DDGS) Using Sieving and Elutriation. Cereal Chemistry 2006;83 324-330.

[130] Singh A. Bioadhesives from Distiller's Dried Grains with Solubles (DDGS) and Studies on Sustainability Issues of Corn Ethanol Industries. Master Thesis, Michigan State University; 2008.

[131] Mohanty AK, Wu Q, Singh A. Bioadhesive from Distillers' Dried Grains with Solubles (DDGS) and the Methods of Making Those. US Patent 2009; US 7618660 B2.

[132] Hu C, Reddy N, Luo Y, Yan K, Yang Y. Thermoplastics from Acetylated Zein-and-Oil-Free Corn Distillers Dried Grains with Solubles. Biomass and Bioenergy 2011;35 884-892.

[133] Hu C, Reddy N, Yan K, Yang Y. Synthesis and Characterization of Highly Flexible Thermoplastic Films from Cyanoethylated Corn Distillers Dried Grains with Solubles. Journal of Agricultural and Food Chemistry 2011;59 1723-1728.

[134] Yu F, Le Z, Chen P, Liu Y, Lin X, Ruan R. Atmospheric Pressure Liquefaction of Dried Distillers Grains (DDG) and Making Polyurethane Foams from Liquefied DDG. Applied Biochemistry and Biotechnology 2008;148 235-243

[135] Youngquist JA, Krzysik AM, English BW, Spelter HN, Chow P. Proceedings of 1996 symposium on The Use of Recycled Wood and Paper in Building Applications, Madison, WI, Forest products society 1996; p 123–134.

[136] Jústiz-Smith NG, Junior Virgo G, Buchanan VE. Potential of Jamaican Banana, Coconut Coir and Bagasse Fibres as Composite Materials. Materials Characterization 2008;59 1273-1278.

[137] Zárate CN, Aranguren MI, Reboredo MM. Resol-Vegetable Fibers Composites. Journal of Applied Polymer Science 2000;77 1832-1840.

[138] de Sousa MV, Monteiro SN, d'Almeida JRM. Evaluation of Pre-Treatment, Size and Molding Pressure on Flexural Mechanical Behavior of Chopped Bagasse–Polyester Composites. Polymer Testing 2004;23 253-258.

[139] Satyanarayana KG, Guimarães JL, Wypych F. Studies on Lignocellulosic Fibers of Brazil. Part I: Source, Production, Morphology, Properties and Applications. Composites Part A: Applied Science and Manufacturing 2007;38 1694-1709.

[140] Shibata S, Cao Y, Fukumoto I. Press Forming of Short Natural Fiber-Reinforced Biodegradable Resin: Effects of Fiber Volume and Length on Flexural Properties. Polymer Testing 2005;24 1005-1011.

[141] Cao Y, Shibata S, Fukumoto I. Mechanical Properties of Biodegradable Composites Reinforced with Bagasse Fibre before and after Alkali Treatments. Composites Part A: Applied Science and Manufacturing 2006;37 423-429.

[142] Vilay V, Mariatti M, Mat Taib R, Todo M. Effect of Fiber Surface Treatment and Fiber Loading on the Properties of Bagasse Fiber–Reinforced Unsaturated Polyester Composites. Composites Science and Technology 2008;68 631-638.

[143] McLaughlin EC. The Strength of Bagasse Fibre-Reinforced Composites. Journal of Materials Science 1980;15 886-890.

[144] Carvajal O, Valdés JL, Puig J. Bagasse Particleboards for Building Purpose. European Journal of Wood and Wood Products 1996;54 61-63.

[145] Aggarwal LK. Bagasse-Reinforced Cement Composites. Cement and Concrete Composites 1995;17 107-112.

[146] Bilba K, Arsene M-A, Ouensanga A. Sugar Cane Bagasse Fibre Reinforced Cement Composites. Part I. Influence of the Botanical Components of Bagasse on the Setting of Bagasse/Cement Composite. Cement and Concrete Composites 2003;25 91-96.

[147] Shen KC. Process for Manufacturing Composite Products from Lignocellulosic Materials. US Patent 1986; US 4627951.

[148] Mobarak F, Fahmy Y, Augustin H. Binderless Lignocellulose Composite from Bagasse and Mechanism of Self-Bonding. Holzforschung 1982;36 131-135.

[149] Patil YP, Gajre B, Dusane D, Chavan S, Mishra S. Effect of Maleic Anhydride Treatment on Steam and Water Absorption of Wood Polymer Composites Prepared from Wheat Straw, Cane Bagasse, and Teak Wood Sawdust Using Novolac as Matrix. Journal of Applied Polymer Science 2000;77 2963-2967.

[150] Paiva JMF, Frollini E. Sugarcane Bagasse Reinforced Phenolic and Lignophenolic Composites. Journal of Applied Polymer Science 2002;83 880-888.

[151] Salyer IO, Usmani AM. Utilization of Bagasse in New Composite Building Materials. Industrial & Engineering Chemistry Product Research and Development 1982;21 17-23.

[152] Stael GC, Tavares MIB, d'Almeida JRM. Evaluation of Sugar Cane Bagasse Waste as Reinforcement in Eva Matrix Composite Materials. Polymer-Plastics Technology and Engineering 2001;40 217-223.

[153] Stael GC, Tavares MIB, d'Almeida JRM. Impact Behavior of Sugarcane Bagasse Waste–EVA Composites. Polymer Testing 2001;20 869-872.

[154] Luz SM, Gonçalves AR, Del'Arco Jr AP. Mechanical Behavior and Microstructural Analysis of Sugarcane Bagasse Fibers Reinforced Polypropylene Composites. Composites Part A: Applied Science and Manufacturing 2007;38 1455-1461.

[155] Mulinari DR, Voorwald HJC, Cioffi MOH, da Silva MLCP, Luz SM. Preparation and Properties of Hdpe/Sugarcane Bagasse Cellulose Composites Obtained for Thermokinetic Mixer. Carbohydrate Polymers 2009;75 317-321.

[156] Chen Y, Ishikawa Y, Maekawa T, Zhang Z. Preparation of Acetylated Starch/Bagasse Fiber Composites by Extrusion. Transactions of the ASAE 2006;49 85-90.

[157] Chen Y, Zhang Z, Ishikawa Y, Maekawa T. Production of Starch-Based Biodegradable Plastics Reinforced with Bagasse Fiber. Journal of the Society of Agricultural Structures 2002;32 177-184.

[158] Hassan ML, Rowell RM, Fadl NA, Yacoub SF, Christainsen AW. Thermoplasticization of Bagasse. I. Preparation and Characterization of Esterified Bagasse Fibers. Journal of Applied Polymer Science 2000;76 561-574.

[159] Hassan ML, Rowell RM, Fadl NA, Yacoub SF, Christainsen AW. Thermoplasticization of Bagasse. II. Dimensional Stability and Mechanical Properties of Esterified Bagasse Composite. Journal of Applied Polymer Science 2000;76 575-586.

[160] Wang J, Manley RSJ, Feldman D. Synthetic Polymer-Lignin Copolymers and Blends. Progress in Polymer Science (UK) 1992;17 611-646.

[161] Lora JH, Glasser WG. Recent Industrial Applications of Lignin: A Sustainable Alternative to Nonrenewable Materials. Journal of Polymers and the Environment 2002;10 39-48.

[162]   Thielemans W, Can E, Morye SS, Wool RP. Novel Applications of Lignin in Compo-
        site Materials. Journal of Applied Polymer Science 2002;83 323-331.

[163]   Rodríguez-Mirasol J, Cordero T, Rodríguez JJ. Preparation and Characterization of
        Activated Carbons from Eucalyptus Kraft Lignin. Carbon 1993;31 87-95.

[164]   Nitz H, Semke H, Mülhaupt R. Influence of Lignin Type on the Mechanical Proper-
        ties of Lignin Based Compounds. Macromolecular Materials and Engineering
        2001;286 737-743.

[165]   Kadla JF, Kubo S. Lignin-Based Polymer Blends: Analysis of Intermolecular Interac-
        tions in Lignin–Synthetic Polymer Blends. Composites Part A: Applied Science and
        Manufacturing 2004;35 395-400.

[166]   Alexy P, Košíková B, Podstránska G. The Effect of Blending Lignin with Polyethy-
        lene and Polypropylene on Physical Properties. Polymer 2000;41 4901-4908.

[167]   Košíková B, Gregorová A. Sulfur Free Lignin as Reinforcing Component of Styrene–
        Butadiene Rubber. Journal of Applied Polymer Science 2005;97 924-929.

[168]   Toriz G, Denes F, Young RA. Lignin Polypropylene Composites. Part 1: Composites
        from Unmodified Lignin and Polypropylene. Polymer Composites 2002;23 806-813.

[169]   Canetti M, Bertini F, De Chirico A, Audisio G. Thermal Degradation Behaviour of
        Isotactic Polypropylene Blended with Lignin. Polymer Degradation and Stability
        2006;91 494-498.

[170]   Mikulášová M, Košíková B. Biodegradability of Lignin—Polypropylene Composite
        Films. Folia Microbiologica 1999;44 669-672.

[171]   El Raghi S, Zahran RR, Gebril BE. Effect of Weathering on Some Properties of Poly-
        vinyl Chloride/Lignin Blends. Materials Letters 2000;46 332-342.

[172]   Banu D, El-Aghoury A, Feldman D. Contributions to Characterization of Poly (Vinyl
        Chloride)–Lignin Blends. Journal of Applied Polymer Science 2006;101 2732-2748.

[173]   Kubo S, Kadla JF. The Formation of Strong Intermolecular Interactions in Immiscible
        Blends of Poly (Vinyl Alcohol)(PVA) and Lignin. Biomacromolecules 2003;4 561-567.

[174]   Fernandes DM, Winkler Hechenleitner AA, Job AE, Radovanocic E, Gómez Pineda
        EA. Thermal and Photochemical Stability of Poly (Vinyl Alcohol)/Modified Lignin
        Blends. Polymer Degradation and Stability 2006;91 1192-1201.

[175]   Kubo S, Kadla JF. Poly (Ethylene Oxide)/Organosolv Lignin Blends: Relationship be-
        tween Thermal Properties, Chemical Structure, and Blend Behavior. Macromolecules
        2004;37 6904-6911.

[176]   Kubo S, Kadla JF. Kraft Lignin/Poly (Ethylene Oxide) Blends: Effect of Lignin Struc-
        ture on Miscibility and Hydrogen Bonding. Journal of Applied Polymer Science
        2005;98 1437-1444.

[177] Camargo FA, Innocentini-Mei LH, Lemes AP, Moraes SG, Durán N. Processing and Characterization of Composites of Poly (3-Hydroxybutyrate-Co-Hydroxyvalerate) and Lignin from Sugar Cane Bagasse. Journal of Composite Materials 2012;46 417-425.

[178] Mousavioun P, Doherty WOS, George G. Thermal Stability and Miscibility of Poly (Hydroxybutyrate) and Soda Lignin Blends. Industrial Crops and Products 2010;32 656-661.

[179] Vengal JC, Srikumar M. Processing and Study of Novel Lignin-Starch and Lignin-Gelatin Biodegradable Polymeric Films. Trends in Biomaterials and Artificial Organs 2005;18 237-241.

[180] Réti C, Casetta M, Duquesne S, Bourbigot S, Delobel R. Flammability Properties of Intumescent PLA Including Starch and Lignin. Polymers for Advanced Technologies 2008;19 628-635.

[181] El Mansouri NE, Yuan Q, Huang F. Characterization of Alkaline Lignins for Use in Phenol-Formaldehyde and Epoxy Resins. BioResources 2011;6 2647-2662.

[182] Peng W, Riedl B. The Chemorheology of Phenol-Formaldehyde Thermoset Resin and Mixtures of the Resin with Lignin Fillers. Polymer 1994;35 1280-1286.

[183] Guigo N, Mija A, Vincent L, Sbirrazzuoli N. Eco-Friendly Composite Resins Based on Renewable Biomass Resources: Polyfurfuryl Alcohol/Lignin Thermosets. European Polymer Journal 2010;46 1016-1023.

[184] Thielemans W, Wool RP. Butyrated Kraft Lignin as Compatibilizing Agent for Natural Fiber Reinforced Thermoset Composites. Composites Part A: Applied Science and Manufacturing 2004;35 327-338.

[185] Nonaka Y, Tomita B, Hatano Y. Synthesis of Lignin/Epoxy Resins in Aqueous Systems and Their Properties. Holzforschung-International Journal of the Biology, Chemistry, Physics and Technology of Wood 2009;51 183-187.

[186] Kumaran MG, De SK. Utilization of Lignins in Rubber Compounding. Journal of Applied Polymer Science 1978;22 1885-1893.

[187] Wang H, Easteal AJ, Edmonds N. Prevulcanized Natural Rubber Latex/Modified Lignin Dispersion for Water Vapour Barrier Coatings on Paperboard Packaging. Advanced Materials Research 2008;47-50 93-96.

[188] Tibenham FJ, Grace NS. Compounding Natural Rubber with Lignin and Humic Acid. Industrial & Engineering Chemistry 1954;46 824-828.

[189] Nakamura K, Hatakeyama T, Hatakeyama H. Thermal Properties of Solvolysis Lignin Derived Polyurethanes. Polymers for Advanced Technologies 1992;3 151-155.

[190] Nakamura K, Mörck R, Reimann A, Kringstad KP, Hatakeyama H. Mechanical Properties of Solvolysis Lignin Derived Polyurethanes. Polymers for Advanced Technologies 1991;2 41-47.

[191] Sarkar S, Adhikari B. Synthesis and Characterization of Lignin–HTPB Copolyur-
ethane. European Polymer Journal 2001;37 1391-1401.

[192] Saraf VP, Glasser WG. Engineering Plastics from Lignin. III. Structure Property Relationships in Solution Cast Polyurethane Films. Journal of Applied Polymer Science
1984;29 1831-1841.

[193] Saraf VP, Glasser WG, Wilkes GL, McGrath JE. Engineering Plastics from Lignin. VI.
Structure–Property Relationships of PEG Containing Polyurethane Networks. Journal of Applied Polymer Science 1985;30 2207-2224.

[194] Saraf VP, Glasser WG, Wilkes GL. Engineering Plastics from Lignin. VII. Structure
Property Relationships of Poly (Butadiene Glycol) Containing Polyurethane Networks. Journal of Applied Polymer Science 1985;30 3809-3823.

[195] Thring RW, Vanderlaan MN, Griffin SL. Polyurethanes from Alcell® Lignin. Biomass
and Bioenergy 1997;13 125-132.

[196] Yoshida H, Mörck R, Kringstad KP, Hatakeyama H. Kraft Lignin in Polyurethanes I.
Mechanical Properties of Polyurethanes from a Kraft Lignin–Polyether Triol–Polymeric MDI System. Journal of Applied Polymer Science 1987;34 1187-1198.

[197] Yoshida H, Mörck R, Kringstad KP, Hatakeyama H. Kraft Lignin in Polyurethanes.
II. Effects of the Molecular Weight of Kraft Lignin on the Properties of Polyurethanes
from a Kraft Lignin–Polyether Triol–Polymeric Mdi System. Journal of Applied Polymer Science 1990;40 1819-1832.

[198] Haars A, Kharazipour A, Zanker H, Hüttermann A. Room-Temperature Curing Adhesives Based on Lignin and Phenoloxidases. In: Hemingway RW, Conner AH, Branham SJ (eds), Adhesives from Renewable Resources, American Chemical Society,
ACS Symposium Series 1989;385 126-134 (DOI: 10.1021/bk-1989-0385.ch010).

[199] El Mansouri N, Pizzi A, Salvadó J. Lignin-Based Wood Panel Adhesives without
Formaldehyde. European Journal of Wood and Wood Products 2007;65 65-70.

[200] Schneider MH, Phillips JG. Furfuryl Alcohol and Lignin Adhesive Composition. US
Patents 2004; US 6747076 B2.

[201] Pizzi A, Cameron F-A, van der Klashorst GH. Soda Bagasse Lignin Adhesive for Particleboard. In: Hemingway RW, Conner AH, Branham SJ (eds), Adhesives from Renewable Resources, American Chemical Society, ACS Symposium Series 1989;385
82-95 (DOI: 10.1021/bk-1989-0385).

[202] Khan MA, Ashraf SM, Malhotra VP. Development and Characterization of a Wood Adhesive Using Bagasse Lignin. International Journal of Adhesion and Adhesives 2004;24 485-493.

[203] Allen MJ, Tung VC, Kaner RB. Honeycomb Carbon: A Review of Graphene. Chemical Reviews 2010;110 132.

[204] Niyogi S, Hamon MA, Hu H, Zhao B, Bhowmik P, Sen R, et al. Chemistry of Single-Walled Carbon Nanotubes. Accounts of Chemical Research 2002;35 1105-1113.

[205] Kim C, Yang KS. Electrochemical Properties of Carbon Nanofiber Web as an Electrode for Supercapacitor Prepared by Electrospinning. Applied Physics Letters 2003;83 1216-1218.

[206] Kim H, Abdala AA, Macosko CW. Graphene/Polymer Nanocomposites. Macromolecules 2010; 43 6515–6530.

[207] White RJ, Tauer K, Antonietti M, Titirici M-M. Functional Hollow Carbon Nanospheres by Latex Templating. Journal of the American Chemical Society 2010, 132, 17360–17363.

[208] Kang X, Wang J, Wu H, Liu J, Aksay IA, Lin Y. A Graphene-Based Electrochemical Sensor for Sensitive Detection of Paracetamol. Talanta 2010;81 754-759.

[209] Suriani AB, Azira AA, Nik SF, Md Nor R, Rusop M. Synthesis of Vertically Aligned Carbon Nanotubes Using Natural Palm Oil as Carbon Precursor. Materials Letters 2009;63 2704-2706.

[210] Sharon M, Sharon M. Carbon Nanomaterials and Their Synthesis from Plant Derived Precursors. Synthesis and Reactivity in Inorganic, Metal-Organic and Nano-Metal Chemistry 2006;36 265-279.

[211] Sharon M, Soga T, Afre R, Sathiyamoorthy D, Dasgupta K, Bhardwaj S, et al. Hydrogen Storage by Carbon Materials Synthesized from Oil Seeds and Fibrous Plant Materials. International Journal of Hydrogen Energy 2007;32 4238-4249.

[212] Yu SH, Cui XJ, Li LL, Li K, Yu B, Antonietti M, et al. From Starch to Metal/Carbon Hybrid Nanostructures: Hydrothermal Metal Catalyzed Carbonization. Advanced Materials 2004;16 1636-1640.

[213] Hayashi J, Kazehaya A, Muroyama K, Watkinson AP. Preparation of Activated Carbon from Lignin by Chemical Activation. Carbon 2000;38 1873-1878.

[214] Kubo S, Kadla JF. Lignin-Based Carbon Fibers: Effect of Synthetic Polymer Blending on Fiber Properties. Journal of Polymers and the Environment 2005;13 97-105.

[215] Babel K, Jurewicz K. KOH Activated Lignin Based Nanostructured Carbon Exhibiting High Hydrogen Electrosorption. Carbon 2008;46 1948-1956.

[216] Gonugunta P. Synthesis and Characterization of Biobased Carbon Nanoparticles from Lignin. Thesis, Master of Applied Science, University of Guelph, 2012.

[217] Gonugunta P, Vivekanandhan S, Mohanty AK, Misra M. A Study on Synthesis and Characterization of Biobased Carbon Nanoparticles from Lignin. World Journal of Nano Science and Engineering 2012;2 148-153.

[218] Ramesh Kumar P, Khan N, Vivekanandhan S, Satyanarayana N, Mohanty AK, Misra M. Nanofibers: Effective Generation by Electrospinning and Their Applications. Journal of Nanoscience and Nanotechnology 2012;12 1-25.

[219] Lallave M, Bedia J, Ruiz-Rosas R, Rodríguez-Mirasol J, Cordero T, Otero JC, et al. Filled and Hollow Carbon Nanofibers by Coaxial Electrospinning of Alcell Lignin without Binder Polymers. Advanced Materials 2007;19 4292-4296.

[220] Schreiber M, Vivekanandhan S, Mohanty AK, Misra M. A Study on the Electrospinning Behaviour and Nanofibre Morphology of Anionically Charged Lignin. Advanced Materials Letters 2012; 3(6), 476-480.

[221] Dallmeyer I, Ko F, Kadla JF. Electrospinning of Technical Lignins for the Production of Fibrous Networks. Journal of Wood Chemistry and Technology 2010;30 315-329.

[222] Seo DK, Jeun JP, Kim HB, Kang PH. Preparation and Characterization of the Carbon Nanofiber Mat Produced from Electrospun Pan/Lignin Precursors by Electron Beam Irradiation. Reviews on Advanced Materials Science 2011;28 31-34.

[223] Spender J, Demers AL, Xie X, Cline AE, Earle MA, Ellis LD, Neivandt DJ, Method for Production of Polymer and Carbon Nanofibers from Water-Soluble Polymers, Nano Letters, 2012; 12 3857-3860.

[224] Kumar R, Choudhary V, Mishra S, Varma IK, Mattiason B. Adhesives and Plastics Based on Soy Protein Products. Industrial Crops and Products 2002;16 155-172.

[225] Shukla R, Cheryan M. Zein: The Industrial Protein from Corn. Industrial Crops and Products 2001;13 171-192.

[226] Reddy M, Mohanty AK, Misra M. Thermoplastics from Soy Protein: A Review on Processing, Blends and Composites. Journal of Biobased Materials and Bioenergy 2010;4 298-316.

[227] Lawton JW, Selling GW, Willett JL. Corn Gluten Meal as a Thermoplastic Resin: Effect of Plasticizers and Water Content. Cereal Chemistry Journal 2008;85 102-108.

[228] di Gioia L, Cuq B, Guilbert S. Effect of Hydrophilic Plasticizers on Thermomechanical Properties of Corn Gluten Meal. Cereal Chemistry Journal 1998;75 514-519.

[229] Di Gioia L, Guilbert S. Corn Protein-Based Thermoplastic Resins: Effect of Some Polar and Amphiphilic Plasticizers. Journal of Agricultural and Food Chemistry 1999;47 1254-1261.

[230] Corradini E, Marconcini JM, Agnelli JAM, Mattoso LHC. Thermoplastic Blends of Corn Gluten Meal/Starch (CGM/Starch) and Corn Gluten Meal/Polyvinyl Alcohol and Corn Gluten Meal/Poly (Hydroxybutyrate-Co-Hydroxyvalerate) (CGM/PHB-V). Carbohydrate Polymers 2011;83 959-965.

[231] Aithani D, Mohanty AK. Value-Added New Materials from Byproduct of Corn Based Ethanol Industries: Blends of Plasticized Corn Gluten Meal and Poly(ε-Caprolactone). Industrial & Engineering Chemistry Research 2006;45 6147-6152.

[232] Wu Q, Sakabe H, Isobe S. Processing and Properties of Low Cost Corn Gluten Meal/Wood Fiber Composite. Industrial & Engineering Chemistry Research 2003;42 6765-6773.

[233] Samarasinghe S, Easteal AJ, Edmonds NR. Biodegradable Plastic Composites from Corn Gluten Meal. Polymer International 2008;57 359-364.

[234] Beg MDH, Pickering KL, Weal SJ. Corn Gluten Meal as a Biodegradable Matrix Material in Wood Fibre Reinforced Composites. Materials Science and Engineering: A 2005;412 7-11.

[235] Diebel W, Reddy MM, Misra M, Mohanty A. Material Property Characterization of Co-Products from Biofuel Industries: Potential Uses in Value-Added Biocomposites. Biomass and Bioenergy, 2012; 37 88–96.

[236] Reddy MM, Mohanty AK, Misra M. Biodegradable Blends from Plasticized Soy Meal, Polycaprolactone, and Poly (Butylene Succinate). Macromolecular Materials and Engineering 2012; 297 455–463.

[237] Reddy MM, Mohanty AK, Misra M. Optimization of Tensile Properties Thermoplastic Blends from Soy and Biodegradable Polyesters: Taguchi Design of Experiments Approach. Journal of Materials Science 2012; 37 2591-2599.

[238] Wu Q, Selke S, Mohanty AK. Processing and Properties of Biobased Blends from Soy Meal and Natural Rubber. Macromolecular Materials and Engineering 2007;292 1149-1157.

[239] Mohanty AK, Wu Q, Selke S. "Green" Materials from Soy Meal and Natural Rubber Blends. US Patents 2010; US 7649036 B2.

[240] Kim H-W, Kim K-M, Ko E-J, Lee S-K, Ha S-D, Song K-B, et al. Development of Antimicrobial Edible Film from Defatted Soybean Meal Fermented by Bacillus Subtilis. Journal of Microbiology and Biotechnology 2004;14 1303-1309.

[241] Kim H-W, Ko E-J, Ha S-D, Song K-B, Park S-K, Chung D-H, et al. Physical, Mechanical, and Antimicrobial Properties of Edible Film Produced from Defatted Soybean Meal Fermented by Bacillus Subtilis. Journal of Microbiology and Biotechnology 2005;15 815-822.

[242] Jiugao Y, Ning W, Xiaofei M. The Effects of Citric Acid on the Properties of Thermo-plastic Starch Plasticized by Glycerol. Starch Stärke 2005;57 494-504.

[243] Fishman ML, Coffin DR, Konstance RP, Onwulata CI. Extrusion of Pectin/Starch Blends Plasticized with Glycerol. Carbohydrate Polymers 2000;41 317-325.

[244] Yu J, Gao J, Lin T. Biodegradable Thermoplastic Starch. Journal of Applied Polymer Science 1996;62 1491-1494.

[245] Wang L, Shogren RL, Carriere C. Preparation and Properties of Thermoplastic Starch Polyester Laminate Sheets by Coextrusion. Polymer Engineering & Science 2000;40 499-506.

[246] Barrett A, Kaletunç G, Rosenburg S, Breslauer K. Effect of Sucrose on the Structure, Mechanical Strength and Thermal Properties of Corn Extrudates. Carbohydrate Poly-mers 1995;26 261-269.

[247] Kim J, Wang N, Chen Y, Yun G-Y. An Electro-Active Paper Actuator Made with Lith-ium Chloride/Cellulose Films: Effects of Glycerol Content and Film Thickness. Smart Materials and Structures 2007;16 1564-1569.

[248] Cervera MF, Heinämäki J, Krogars K, Jörgensen AC, Karjalainen M, Colarte AI, et al. Solid-State and Mechanical Properties of Aqueous Chitosan-Amylose Starch Films Plasticized with Polyols. AAPS Pharmscitech 2004;5 109-114.

[249] Bergo P, Sobral PJA. Effects of Plasticizer on Physical Properties of Pigskin Gelatin Films. Food Hydrocolloids 2007;21 1285-1289.

[250] Clarizio SC, Tatara RA. Tensile Strength, Elongation, Hardness, and Tensile and Flexural Moduli of PLA Filled with Glycerol-Plasticized DDGS. Journal of Polymers and the Environment, 2012; 20 638-646.

[251] Galietta G, Di Gioia L, Guilbert S, Cuq B. Mechanical and Thermomechanical Proper-ties of Films Based on Whey Proteins as Affected by Plasticizer and Crosslinking Agents. Journal of Dairy Science 1998;81 3123-3130.

[252] Rosa DS, Bardi MAG, Machado LDB, Dias DB, Andrade e Silva LG, Kodama Y. Influ-ence of Thermoplastic Starch Plasticized with Biodiesel Glycerol on Thermal Proper-ties of PP Blends. Journal of Thermal Analysis and Calorimetry 2009;97 565-570.

[253] Rosa DS, Bardi MAG, Machado LDB, Dias D, Silva LGA, Kodama Y. Starch Plasti-cized with Glycerol from Biodiesel and Polypropylene Blends. Journal of Thermal Analysis and Calorimetry 2010;102 181-186.

[254] Dakka JM, Mozeleski EJ, Baugh LS, Process for making triglyceride plasticizer from crude glycerol, US Patent 2012; US 8,299,281 B2.

[255] Nobrega MM, Olivato JB, Bilck AP, Grossmann MVE, Yamashita F. Glycerol with Different Purity Grades Derived from Biodiesel: Effect on the Mechanical and Viscoe-

lastic Properties of Biodegradable Strands and Films. Materials Science and Engineering: C 2012; 32 2220-2222.

[256] Grazhulevichene V, Augulis L, Gumbite M, Makarevichene V. Biodegradable Composites of Polyvinyl Alcohol and Coproducts of Diesel Biofuel Production. Russian Journal of Applied Chemistry 2011;84 719-723.

[257] Stevens ES, Ashby RD, Solaiman DKY. Gelatin Plasticized with a Biodiesel Coproduct Stream. Journal of Biobased Materials and Bioenergy 2009;3 57-61.

[258] Bardi MAG, Machado LDB, Guedes CGF, Rosa DS. Influence of Processing on Properties of Aditived Polypropylene, Starch Plasticized with Glycerol from Biodiesel and Their Blends. http://www.ipen.br/biblioteca/cd/cbpol/2009/PDF/1450.pdf. (Accessed on 3rd December 2012)

[259] Singh M, Milano J, Stevens ES, Ashby RD, Solaiman DKY. Gelatin Films Plasticized with a Simulated Biodiesel Coproduct Stream. eXPRESS Polymer Letters; 2009; 3 201-206.

[260] de Moura CVR, Nunes ASL, Neto JMM, Neres HLS, de Carvalho LMG, de Moura EM. Synthesis and Characterization of Polyesters from Glycerol by-Product of Biodiesel Production. Journal of the Brazilian Chemical Society 2012;23 1226-1231.

[261] Brioude MM, Guimarães DH, Fiúza RP, Prado LASA, Boaventura JS, José NM. Synthesis and Characterization of Aliphatic Polyesters from Glycerol, by-Product of Biodiesel Production, and Adipic Acid. Materials Research 2007;10 335-339.

[262] Carnahan MA, Grinstaff MW. Synthesis of Generational Polyester Dendrimers Derived from Glycerol and Succinic or Adipic Acid. Macromolecules 2006;39 609-616.

[263] Tang J, Zhang Z, Song Z, Chen L, Hou X, Yao K. Synthesis and Characterization of Elastic Aliphatic Polyesters from Sebacic Acid, Glycol and Glycerol. European Polymer Journal 2006;42 3360-3366.

[264] Holser RA, Willett JL, Vaughn SF. Thermal and Physical Characterization of Glycerol Polyesters. Journal of Biobased Materials and Bioenergy 2008;2 94-96.

[265] Stumbé J-F, Bruchmann B. Hyperbranched Polyesters Based on Adipic Acid and Glycerol. Macromolecular Rapid Communications 2004;25 921-924.

[266] Nicol RW, Marchand K, Lubitz WD. Bioconversion of Crude Glycerol by Fungi. Applied Microbiology and Biotechnology 2012; 93 1865-1875.

[267] Johnson DT, Taconi KA. The Glycerin Glut: Options for the Value Added Conversion of Crude Glycerol Resulting from Biodiesel Production. Environmental Progress 2007;26 338-348.

Permissions

The contributors of this book come from diverse backgrounds, making this book a truly international effort. This book will bring forth new frontiers with its revolutionizing research information and detailed analysis of the nascent developments around the world.

We would like to thank Zhen Fang, for lending his expertise to make the book truly unique. He has played a crucial role in the development of this book. Without his invaluable contribution this book wouldn't have been possible. He has made vital efforts to compile up to date information on the varied aspects of this subject to make this book a valuable addition to the collection of many professionals and students.

This book was conceptualized with the vision of imparting up-to-date information and advanced data in this field. To ensure the same, a matchless editorial board was set up. Every individual on the board went through rigorous rounds of assessment to prove their worth. After which they invested a large part of their time researching and compiling the most relevant data for our readers. Conferences and sessions were held from time to time between the editorial board and the contributing authors to present the data in the most comprehensible form. The editorial team has worked tirelessly to provide valuable and valid information to help people across the globe.

Every chapter published in this book has been scrutinized by our experts. Their significance has been extensively debated. The topics covered herein carry significant findings which will fuel the growth of the discipline. They may even be implemented as practical applications or may be referred to as a beginning point for another development. Chapters in this book were first published by InTech; hereby published with permission under the Creative Commons Attribution License or equivalent.

The editorial board has been involved in producing this book since its inception. They have spent rigorous hours researching and exploring the diverse topics which have resulted in the successful publishing of this book. They have passed on their knowledge of decades through this book. To expedite this challenging task, the publisher supported the team at every step. A small team of assistant editors was also appointed to further simplify the editing procedure and attain best results for the readers.

Our editorial team has been hand-picked from every corner of the world. Their multi-ethnicity adds dynamic inputs to the discussions which result in innovative outcomes. These outcomes are then further discussed with the researchers and contributors who give their valuable feedback and opinion regarding the same. The feedback is then collaborated with the researches and they are edited in a comprehensive manner to aid the understanding of the subject.

Apart from the editorial board, the designing team has also invested a significant amount of their time in understanding the subject and creating the most relevant covers. They scrutinized every image to scout for the most suitable representation of the subject and create an appropriate cover for the book.

The publishing team has been involved in this book since its early stages. They were actively engaged in every process, be it collecting the data, connecting with the contributors or procuring relevant information. The team has been an ardent support to the editorial, designing and production team. Their endless efforts to recruit the best for this project, has resulted in the accomplishment of this book. They are a veteran in the field of academics and their pool of knowledge is as vast as their experience in printing. Their expertise and guidance has proved useful at every step. Their uncompromising quality standards have made this book an exceptional effort. Their encouragement from time to time has been an inspiration for everyone.

The publisher and the editorial board hope that this book will prove to be a valuable piece of knowledge for researchers, students, practitioners and scholars across the globe.

List of Contributors

Robert Diltz
Air Force Research Laboratory, Tyndall AFB, FL, USA

Pratap Pullammanappallil
Department of Agricultural and Biological Engineering, University of Florida, Gainesville, FL, USA

Emad A. Shalaby
Biochemistry Dept., Faculty of Agriculture, University of Ruyku, Cairo University, Egypt

Fung Min Liew, Michael Köpke and Séan Dennis Simpson
LanzaTech NZ Ltd., Parnell, Auckland, New Zealand

Anli Geng
School of Life Sciences and Chemical Technology, Ngee Ann Polytechnic, Singapore

Valeriy Chernyak, Oleg Nedybaliuk, Sergei Sidoruk, Vitalij Yukhymenko, Eugen Martysh, Olena Solomenko and Yulia Veremij
Taras Shevchenko National University of Kyiv, Ukraine

Alexandr Tsimbaliuk
Institute of Physics, National Academy of Sciences of Ukraine, Kyiv, Ukraine

Dmitry Levko
Institute of Physics, National Academy of Sciences of Ukraine, Kyiv, Ukraine
Physics Department, Technion, 32000, Haifa, Israel

Leonid Simonchik and Andrej Kirilov
B.I. Stepanov Institute of Physics, National Academy of Sciences, Minsk, Belorus

Oleg Fedorovich
Institute of Nuclear Research, National Academy of Sciences of Ukraine, Kyiv, Ukraine

Anatolij Liptuga
V.E.Lashkaryov Institute of Semiconductor Physics, National Academy of Science of Ukraine, Kyiv, Ukraine

Valentina Demchina
The Gas Institute, National Academy of Science of Ukraine, Kyiv, Ukraine

Semen Dragnev
National University of Life and Environmental Sciences of Ukraine, Kyiv, Ukraine

Hongjuan Liu, Genyu Wang and Jianan Zhang
Institute of Nuclear and New Energy Technology, Tsinghua University, Beijing, P. R., China

Edilson León Moreno Cárdenas, Deisy Juliana Cano Quintero and Cortés Marín Elkin Alonso
Engineering Agricultural Department, National University of Colombia, Medellín, Colombia

S. Vivekanandhan, N. Zarrinbakhsh, M. Misra and A. K. Mohanty
Bioproducts Discovery and Development Centre, Department of Plant Agriculture, Crop Science Building, University of Guelph, Guelph, ON, N1G 2W1, Canada
School of Engineering, Thornbrough Building, University of Guelph, Guelph, ON, N1G 2W1, Canada

Printed in the USA
CPSIA information can be obtained
at www.ICGtesting.com
JSHW011432221024
72173JS00004B/766